ECOPEDAGOGIES

Ecopedagogies showcases a range of creative approaches that educators across multiple disciplines use to empower students to access and engage with nature, an increasingly important consideration in a post-COVID world in environmental crisis.

The volume includes chapters written by scholars from the environmental arts and humanities, literature, writing studies, rhetoric, music, religious studies, environmental studies and sustainability, sociology and anthropology, physical education, and outdoor education. Each author walks the reader through the details of how their ecopedagogy works, identifies potential challenges while also detailing how to address them, and explains the rewards to students, instructors, and more-than-human nature that they have witnessed through the use of these approaches. The contributions represent diverse types of academic institutions, offering broad applicability to instructors, including community colleges, private liberal arts colleges, and large state, regional, public, and private universities. The book explores a series of key questions about how educators can facilitate meaningful learning experiences with the natural world, inside and outside the classroom, and it looks at how to foster inclusivity, navigate problems with access, and explore intersections with environmental justice.

As a practical guide, the book delivers a well-provisioned toolbox containing exercises, activity guides, and assignments for those teaching environmentally focused college courses.

Ellen Bayer is an Associate Professor of Environmental Arts and Humanities at the University of Washington Tacoma, USA.

Judson Byrd Finley is Associate Professor and Head of the Department of Sociology and Anthropology at Utah State University, USA.

"The digital pivot in post-pandemic education refocuses the perennial problem of 'getting students out of the classroom and into the wild.' As the classroom morphs in ways that ignore habitat—both the inner and the outer wilds—we need models, leaders, and guides to remind us to pay attention to where we are, and to learn how to be here with others—human and more-than-human. Bayer and Finley's clear prose and diverse roster of contributors provides helpful and inspiring guidance for both seasoned guides and for those taking their first steps into the field."

David Greenwood, *Lakehead University, Canada*

"*Ecopedagogies* provides a diverse set of humanistic and interdisciplinary approaches to place-based experiential education within colleges and universities, which highlight the humane promise of learning with and from other species, as well as the land itself. A resourceful tool for those attempting to explore the *terra nullius* of contemporary ecoliteracy."

Richard Kahn, *Antioch University, USA*

"Rich in both practical advice and wise reflection, this volume is a treasure trove of ideas and encouragement for all who seek to embed learning within the wild community of life. I came away from reading both inspired and invigorated."

David George Haskell, *The University of the South, USA*

"[This] is the book I wish I'd had two decades ago when I was an English teacher at a community college, and a science colleague and I began taking students on field trips. This wide-ranging collection provides philosophical grounding for getting students out of the classroom, offers a variety of curricular approaches ranging from A(laska) to Z(oos), and is packed with practical advice, such as 'define field loosely' and 'meet your students where they are.' Importantly, it addresses head-on questions of inclusion, access, and privilege, pointing out that place-based pedagogies that center student experience through reflection connect students with place *and* with each other, and what it means to me 'in relation.' This timely collection ends with a chapter titled 'Cultivating a Restorative Educational Ethic in a World on Fire,' reminding us that the stakes are high and the time is short. This is an indispensable field guide for teaching in the 21st century."

Holly J. Hughes, *Co-editor of Contemplative Approaches to Sustainability in Higher Education: Theory & Practice (Routledge, 2017)*

ECOPEDAGOGIES

Practical Approaches to Experiential Learning

Edited by Ellen Bayer and Judson Byrd Finley

Routledge
Taylor & Francis Group
LONDON AND NEW YORK

earthscan
from Routledge

Cover image: © The cook tent at night, Black Canyon of the Gunnison National Park; Andrew McAllister (Utah State University Caine College of the Arts)

First published 2023
by Routledge
4 Park Square, Milton Park, Abingdon, Oxon OX14 4RN

and by Routledge
605 Third Avenue, New York, NY 10158

Routledge is an imprint of the Taylor & Francis Group, an informa business

British Library Cataloguing-in-Publication Data
A catalogue record for this book is available from the British Library

Library of Congress Cataloging-in-Publication Data
A catalog record has been requested for this book

ISBN: 978-1-032-11844-4 (hbk)
ISBN: 978-1-032-11845-1 (pbk)
ISBN: 978-1-003-22180-7 (ebk)

DOI: 10.4324/9781003221807

Typeset in Bembo
by Taylor & Francis Books

For my professors who opened the door, and for the students who venture out with me.
For Marcel Kornfeld and Mary Lou Larson (1954–2022), who have created the space that gives so many students the freedom to roam.

CONTENTS

ILLUSTRATIONS

Figures

Table

CONTRIBUTORS

Ellen Bayer is an Associate Professor of Environmental Arts & Humanities at the University of Washington Tacoma, where she teaches literature courses using interdisciplinary, place-based, and experiential approaches. She is currently writing a wilderness memoir that explores her personal journey into the backcountry.

Damiano Benvegnù, PhD, is a Senior Lecturer at Dartmouth College; an Associate Fellow of the Oxford Centre for Animal Ethics; and the Art and Creative Writing Editor for the *European Journal for Literature, Culture, and the Environment.* Benvegnù is the author of *Animals and Animality in Primo Levi's Work.*

Allison Lee Blyler teaches in the College of Arts and Sciences Writing Program, Boston University, USA, with Holly Connell Schaaf (co-author of Chapter 5). Blyler and Schaaf also co-direct Boston Now, an initiative focusing on experiential and place-based learning. Their chapter draws on a seminar they have been co-teaching since Fall 2017, which engages deeply with a local wildlife sanctuary.

Kelly M. Branam Macauley is Director of Global Studies and Professor of Anthropology at St. Cloud State University. She has continuously worked to build bridges between archaeologists and native communities in terms of cultural resource management including co-directing anthropology field schools in Bighorn Canyon National Recreation Area.

Jeremy Chow is an Assistant Professor of English at Bucknell University. His research and teaching explore the environmental humanities, especially attentive to nonhumanity, through literature and media. His first book, *The Queerness of Water: Violent Entanglements in Troubled Ecologies*, is forthcoming from University of Virginia Press.

Lachell Faure is an English teacher for the Social Science and Humanities department of ENSTA Bretagne. She obtained a Master of Arts in international affairs and TESOL certification from Florida State University and has 15+ years teaching experience in the United States, Japan, and France.

Chris Finley is a retired National Park Service archaeologist whose career specialized in historical restoration of historic ranch buildings and collaborative archaeological education and training with Indigenous communities. He lives on a small homestead in rural northwestern Wyoming with a menagerie of horses, cats, and magpies.

Judson Byrd Finley is Associate Professor and Head of the Department of Sociology and Anthropology at Utah State University. Judson is an archaeologist who works at the intersection of culture history, earth science, and environmental change. His current research examines the socioecological history of Utah's first Indigenous farmers.

Levi Gardner is an educator, grower, artist, and independent scholar in Grand Rapids, MI. His work includes developing a university farm, founding a non-profit, and a decade of teaching environmental studies. He has his MS in community sustainability from Michigan State University.

Summer Harrison is Associate Professor of English and Environmental Studies at Drew University, where she teaches courses in environmental humanities, community-based learning, and American literature. Her work examines the intersection of stories, environments, and ecopedagogies.

Amanda Hayden is a Professor at Sinclair College, teaching courses in Religion, Environmental Ethics, and Humanities and is also a Poet Laureate. She has received several Teaching Excellence Awards and was featured in *Ohio Magazine* for outstanding experiential teaching. Hayden recently wrote a children's book about her family's animal rescues at their sustainability-minded farm.

Keri Holt is an associate professor of English at Utah State University and director of the American studies program. She is the author of *Reading These United States: Federal Literacy in the Early United States* (2019) and a co-editor of *Mapping Region in Early American Writing* (2015).

Mary A. Jackson is Associate Faculty at Prescott College. She is a transdisciplinary researcher, former outdoor educator, and artist, with an emphasis in environmental humanities, sustainability, and new materialist methodologies. Her work examines more-than-human agency and place through an embodied awareness of place, confronting both anthropogenic and colonizing narratives.

Davy Knittle is an HMEI/Princeton Mellon Fellow in Architecture, Urbanism & the Environment at Princeton University. He received his PhD in English from the University of Pennsylvania in 2021. His critical work has appeared in *ISLE: Interdisciplinary Studies in Literature and Environment, GLQ: A Journal of Lesbian and Gay Studies*, and *Planning Perspectives*.

Ian MacKenzie teaches environmental writing and rhetoric at Dawson College in Montréal, Canada. He is the Project Lead for Learning Communities and the founding Director of Writing in the Disciplines at Dawson. Running, hiking, paddling, skiing, and cabin loafing are Ian's seasonal preoccupations.

Kevin Maier is Professor of English at the University of Alaska Southeast, where he teaches classes in writing, American literature, and the environmental humanities. His current research explores the intersections of climate politics and outdoor recreation. Outside of work, he tries to keep up with his two boys in the mountains, forests, and waters surrounding his home, which sits on the land of the Áak'w Kwáan Tlingit.

Andrew Niess is an audiovisual researcher and instrument maker. He received a PhD in music studies from the University of Pennsylvania and was a Mellon Dissertation Fellow at the Penn Program in Environmental Humanities. His research is forthcoming in the *Oxford Handbook of Protest Music* and the *Journal of Japanese Studies*.

Darren Paisley is an English Language teacher at ENSTA-Bretagne, a French graduate and postgraduate engineering school based in the city of Brest. He holds a Bachelor of Arts degree along with professional qualifications in teaching and in Aquaculture. He is interested in combining language teaching with social and environmental awareness.

Lauren E. Perry earned her PhD in English from the University of New Mexico and her Master of Arts in English from the University of Wyoming. She also studied at the University of London and at Stirling University. Her scholarship focuses on animals in American environmental literature from the nineteenth and twentieth centuries.

Matthew J. Rowe, PhD, is an Assistant Professor of Practice and the Director of Undergraduate Studies in the School of Anthropology at the University of Arizona. Dr. Rowe is currently working in Northern Arizona on issues of collector-collaboration in archaeological research, and cultural and environmental justice.

Holly Connell Schaaf teaches in the College of Arts and Sciences Writing Program, Boston University, USA, with Allison Lee Blyler (co-author of Chapter 5). Blyler and Schaaf also co-direct Boston Now, an initiative focusing on experiential and place-based learning. Their chapter draws on a seminar they have been co-teaching since Fall 2017, which engages deeply with a local wildlife sanctuary.

Doug Smyth is crazy about cycling and cross-country skiing, and even crazier about bird and wildlife photography. He teaches in the Physical Education Department at Dawson College in Montreal, Canada, where he has served multiple terms as Coordinator of Outdoor Education Programming.

Brad Tabas is Associate Professor in the Department of Human and Social Sciences at ENSTA Bretagne. He earned his BA from the University of Pennsylvania and his MA and PhD at New York University. He his work on education and the environment has appeared in *Ecozon@, Ecocene*, and *Resilience*.

Hubert Burdick Two Leggins is an enrolled member of the Crow Indian Nation. He is currently a cultural counselor for Apsáalooke Healing, a recovery center focused on supporting collaborative and harmonious health for their clients. Hubert spent many years serving on the Crow Indian Nation's culture committee and from 2009 to 2012 served as the Crow Tribal Historic Preservation Officer.

ACKNOWLEDGMENTS

We are grateful to the managing editors and editorial assistants at Routledge who enthusiastically shepherded this project throughout the entire process: Rebecca Brennan, Grace Harrison, Rosie Anderson, and Matthew Shobbrook. We appreciate their belief in our volume, as well as their guidance and support as we worked to bring it to life. We also appreciate the thoughtful feedback provided by the anonymous reviewers of our proposal, which helped us move the project forward.

Thank you to Keri Holt for sharing her valuable feedback and insights on our draft of the introduction.

Andrew McAllister (USU Caine College of the Arts) took the cover photograph.

Ellen would like to thank the Association for the Study of Literature and Environment (ASLE) for giving her the opportunity to propose and chair two panels on experiential learning at its 2018 conference, which served as the seed for this volume. She would also like to express her gratitude to the School of Interdisciplinary Arts and Sciences at the University of Washington Tacoma for awarding her a sabbatical and a Scholarship and Teaching Grant, which provided much needed time and funding for the development of this project. She also thanks her colleague Randy Nichols for helping her to navigate the new-to-her waters of preparing an edited collection. She's thankful for the companionship of Max, Sparkimus, and Buster, who reminded her throughout this process that bird watching, cat naps, and lap snuggles are important components of a healthy work/life balance. Above all, Ellen is eternally grateful to Judson Finley for his partnership in this endeavor. The experience, vision, and passion that he brought to this project made it better than she could have ever imagined.

Judson thanks Keri Holt for partnering in the Utah State University Honors College Think Tank Interdisciplinary Colloquium that led to the work here. Evelyn Funda suggested the journal *Interdisciplinary Studies in Literature and Environment* as a publication outlet for an essay that satisfied the obligation of a USU Mountain West

Center for Regional Studies grant for funding our class trip to Dinosaur National Monument. Without that recommendation, he wouldn't have connected with Ellen's project. Judson is also grateful to Ellen for allowing him to invite himself along on her trip through the development and production of an edited volume. Oddly enough, a global pandemic and many Zoom calls provided the focus that made this work seamless and possible. Judson works in and out of the field with the continuing support of Dr. Maureen Patrice Boyle, partner extraordinaire.

For supplemental information supporting this volume, including course syllabi, lesson plans, class projects, field trip guides, and risk management forms, please visit https://digitalcommons.usu.edu/ecopedagogies.

1

OUT OF THE CLASSROOM AND INTO THE WILD

Ecopedagogies in Action

Ellen Bayer and Judson Byrd Finley

We are here because our teachers sent us out of the classroom.

For Ellen, excursions into the field as an undergraduate learner took many forms. Class trips to the symphony, the theatre, art and history museums, and a farm that was a site of human enslavement helped me to engage with the corresponding literary texts in meaningful and personal ways. That sensory, physical connection of seeing and touching and hearing for myself reached someplace deeper within me than any lecture ever had. The culminating experience came in an interdisciplinary Honors seminar during my final semester, called "Don't Fence Me In." The course examined natural history and literary, film, and musical texts from a place-based perspective, and it required us to go outside—a lot.

I was 22 years old and barely keeping my head above water. I worked two jobs, commuted an hour each way to campus five days per week, and was a full-time student trying desperately to maintain an impressive GPA and catch the attention of top graduate programs. To cap off all the stress, I developed an ulcer that semester as well. When my seminar professor assigned us to choose a place outdoors to visit each week of the term, I had no idea how I would manage to do so. Growing up, I was a child of the forest and spent every possible minute outdoors, roaming wildly. Life felt too busy for that now; my days were a blur of rubber on concrete and windowless interiors. Sitting under a tree in the park was an impossible luxury, not part of a rigorous college course. I would have to convince myself that this was a legitimate use of my sparse time, that this was *homework*.

Wrapping my mind around that was initially difficult, but after the first visit to "my place," a small creek in a wooded area near my mom's suburban home, I realized that Professor Jodi Ferner was on to something. We were reading Annie Dillard's *Pilgrim at Tinker Creek* at the time, and I tried to emulate Dillard's talent for keen observation, wonder, and curiosity as I sat cross-legged by the creekside. I traced my fingers through the dirt and held crumbled leaves to my nose, inhaling

DOI: 10.4324/9781003221807-1

the earthy remains of autumn. I sketched the deciduous trees as they marched into spring, held a vial of creek water to the light and imagined its microscopic world, and tried to identify the birds rummaging through the woodsy detritus. My geology textbook sorted out limestone from shale. I asked questions about the wood's inhabitants and its history, both human and more-than-human. I laid straight down on the ground, closed my eyes, and allowed the robin's chirp to quiet the cacophony in my mind. Being in this place, Dillard's words (Dillard 1974, p. 181) resonated loudly: "Perhaps I don't need a lobotomy, but I could use some calming down, and the creek is just the place for it. I must go down to the creek again." The assignment became a brief respite from the hurry of my daily existence. It was what I needed, more than anything, at that juncture. It rekindled my relationship to the natural world and invited me to continue nurturing a core part of myself that I had neglected. That creek is thousands of miles and over 20 years away from me now, but its dirt remains under my fingernails; its scent lingers in my nose.

Judson came by way of a different path. Growing up in Wyoming, the outdoors was literally out the front door, but scouting was a way to truly get into the mountains. With the lofty post-secondary goal of being a construction worker, I spent time as an unskilled laborer in a northern Nevada goldmine and roofing houses in Wyoming. I was quick to realize that was no life for me, and university was on my immediate horizon. My father and I visited a University of Wyoming archaeology crew in the mountains south of Yellowstone National Park where we camped, worked, and learned about the past for several days. Archaeology became the project that blended the labor of construction with the academic inquiry of the Indigenous past in 10-day hitches camping in the mountains and desert. I never looked back.

Throughout my undergraduate career, I assisted my mentors in all things related to the field, from meal plans and grocery shopping to database design to, when necessary, remote vehicle repairs literally with bubblegum and bailing wire. As a first-year graduate student my goal as part of my MA research was to develop and run my own archaeological field school in Wyoming's Bighorn Mountains. While the sites and research changed, the crews grew, involving students from Indiana, Tennessee, Minnesota, and Wyoming, and, eventually, Crow and Northern Cheyenne Tribal members from southwest Montana. Teaching students how to camp became as important as the archaeology itself. After all, if you can spend six weeks in a tent in Wyoming, you can do anything.

But what really emerged from the field schools was the realization that those experiences bind us to place. The collective action of "the project" creates a common cause. "Mountain Time," is more than a zone. It is a liminal place where the world outside stands still and the work and relationships with your peers, who often become close friends, future colleagues, sometimes life partners, or maybe bitter adversaries take precedent over all else. It is the realization that landscapes are encoded with millennia of human experience that we learn about and add to along the way. The more-than-human cannot be denied or ignored. The mountain and the wind will have their way, as will the bear, the lion, and the wolf. Naming these experiences and positioning them within pedagogies of place became as much a part of the curriculum as archaeological method and camping skills. Creating a context where students can safely share these experiences has come to define my work as a mentor as much as the intellectual discipline that I am situated in now 30 years after Marcel Kornfeld and Mary Lou Larson first took me into their camp.

The experiences our mentors offered us, sending us *out* of the classroom, shape the way we teach our own students today. We suspect the same is true of each contributor to this collection. We know, first-hand and from our students' perspective, how being sent outside can impact learners. We bring that experience to our classrooms, and to this volume, because we believe in the work that experiential and place-based learning does.

Positioning Ecopedagogies as Practical Experiential Learning

The essays we have assembled are positioned within the triad of critical pedagogy of place (Gruenewald 2003), ecopedagogy (Kahn 2008), and experiential learning (Kolb and Kolb 2017; Lewis and Williams 1994). The collection is, by design and through editorial instruction, not a deep dive into the pedagogy of any one of these areas. Neither of us identifies as pedagogues—we are a literary critic and an archaeologist. That being said, we feel it is important to highlight the key elements of each approach that we see here as common threads that knit the collection together. Anyone schooled especially in place-based pedagogy will know it to be a vast literature whose position and relevance have been critically debated over the last two decades. We consider how this literature, intentional or not, informs our contributors' teaching objectives and learning outcomes. The "field," loosely defined among our contributors as anywhere we can take our students outside of the traditional classroom, is the medium that shapes these outcomes. We see no need to overcomplicate the outcomes: the practical approaches our colleagues advocate here cultivate a sense of place, sense of community, and sense of self. Each teacher speaks to basic human emotions through shared experiences they create for their students. Each curriculum is as unique as the teacher who developed it and is designed to spark awareness, and perhaps passion, in every student. In doing so, this collection fills a gap in the critical pedagogy of place by addressing the diverse stakes and stakeholders and the complex interactions between them, which are crucial components of environmental awareness. It enables the activist agenda of ecopedagogies by investing students in the future of conservation movements, which are singular and situational to their community. As one of our contributors notes, long after classroom lessons are forgotten, students will remember how the classroom experience, or the field in our case, *made them feel*. Experiences matter in enabling the guiding pedagogy of place-based education. In the following section, we highlight how our contributors use practical approaches to fulfill the objectives of a critical pedagogy of place and an activist ecopedagogy through experiential learning—the simple act of learning by doing (Kolb and Kolb 2017; Lewis and Williams 1994).

Sense of Place Matters

Keith Basso's (1996) essays collected as *Wisdom Sits in Places* is one bookend of our collection that provides important philosophical framing. Basso's ethnographic work with the Western Apache features a rich world of toponyms that link past, present, and future as lived history. Western Apache tribal members take instruction from the land through constant reminders of the lessons that are encoded in its stories. For Basso (1996, p. 5), "remembering often provides a basis for imagining." Place-making is an active, ongoing process that adds individual experiences to the collective imagining of place-worlds. "If place-making is a way of constructing the past, a venerable means of *doing* history, it is also a way of constructing social traditions, and, in the process, personal and social identities. We *are* the place worlds we imagine (Basso 1996, p. 7; emphasis in original)." The pedagogical message is clear: the experiences we provide our students in the field, however we choose to define it, are a first-order process in creating social traditions that become central to individual identity. Basso's (1996, p. 34) lesson resonates through this collection: "Knowledge of places is therefore closely linked to knowledge of self." This is the central appeal of place-based pedagogy. But as we know, and as the contributors to this collection remind us, the world is more-than-human—stone, claw, and feather, concrete, glass, and steel have equal agency in the place-worlds we inhabit (TallBear 2015, Kimmerer 2017).

Place-making as Pedagogy

The mindfulness of place-based pedagogies that many of our contributors advocate shifts students' awareness from living in places to being in-relation with places, which we see as one of the key contributions of a critical pedagogy of place (Gruenewald 2003). Gruenewald's (2003, pp. 6–8) central concern and a key outcome of his approach was individual and collective well-being while simultaneously advocating for environmental conservation. While some might take issue with his challenge for educators and students to "reinhabit their places" (ibid., p. 7), the message is consistent with Basso's notion of place-making. *We are the place-worlds we imagine.* The key elements of what Gruenewald (ibid., p. 7) terms an ecological place-based pedagogy build on the singularity of place through multidisciplinary approaches to teaching and learning that center student experience through reflection that connects people with place and each other. *This is what it means to be in-relation.* Imagine each essay in this volume as a tributary to the master stream that fills an ocean where stories of place produced through creative, practical teaching are the floating islands of shared experience and collective well-being.

Thus, we position our collection as a set of methods that are easily adaptable to diverse contexts where place-making is the central goal and cultivation of sense of place, self, and community is a primary outcome. In other words, the contributions offered in this volume enable the principles of a critical pedagogy of place. They collectively respond to its call for the coupling of content and skills with student engagement in place. The methods they advocate are transformative and true to the ecopedagogy agenda where resistance can be active protest or passive contemplation. Each contributor to this collection is mindful that their communities are sociopolitical constructs that often marginalize and disenfranchise community members who are Black, Indigenous, and People of Color. We are reminded of this through student messages of differential access to food through campus housing compared with their family homes or the frequency with which trash is collected from city streets. Attention to soundscapes and recycling practices exposes students' position in consumptive culture while raising awareness that the natural world persists in urban centers despite our society's best efforts to pave it over. The breadth of approaches presented here reflects a decadal and generational shift in place-based education that is reactive and reflective. As Levi Gardner notes in his closing essay, it is not easy for us or our students to embrace the mess we're in, but we are confident the lessons our contributors offer respond to Gruenewald's (2003, p. 7) call to "honor learners' developmental readiness for engaging with complex ecological themes." But let us not forget that we can also reduce the lessons of this volume to just good fun. Take your students for a walk in the park. Amend the beds in your campus or community garden. Slog through the mud with an overloaded backpack. Tell stories. Make stories. These are the experiences your students will remember. Ecopedagogies are about place-making that embraces the complex place-world we inhabit and recognizes its power is more-than-human.

Ecopedagogies in Action

We would be remiss not to address the explicitly activist undertones of ecopedagogy as a logical outcome of a critical pedagogy of place (Kahn 2008). Bowers's (2008) critique is based on the fundamental notion that a critical pedagogy of place fails to identify the problem of the commons in conservation movements. Anthropology's central ideal of cultural relativism challenges how we as a complex society of privileged and disaffected communities find value in place. Beauty is truly in the eye of the beholder. While I might seek the solace of a desert night, the brightly lit hydraulic fracking rig on the skyline speaks to my high school best friend, the directional driller. This is another way of illuminating Bowers's call for a more sophisticated engagement with the complexities of community perspectives and interests in a critical pedagogy of place. This collection prioritizes the importance of recognizing and examining the many stakes and stakeholders involved when working with places, and we bookend our framing of this collection by linking the activist ecopedagogy agenda with a useful approach for cultivating activism that underwrites each contribution. Gary Machlis and Jonathan Jarvis offer practical "Strategies for the Future of Conservation" that are operationalized throughout this collection (Machlis and Jarvis 2018, pp. 43–67). Principal activities include organization through collective action and coalitions of common interest, operating with strategic intent, connecting citizens to nature through first-hand experience, knowing and sharing the full American story, engaging locally and with local values, and transferring knowledge and power to the next generation. Every teacher in this collection has designed classes, exercises, and field experiences that speak to each of these actions. Thus, we offer this collection as a set of methods in place-making that fulfills the objectives of a critical pedagogy of place and provides waypoints in a chart for an uncertain future.

Lessons Learned in the Field

Our contributors Ian MacKenzie and Doug Smyth share their three-ingredient recipe for a successful trip to the field:

1. strenuous and intellectually challenging outdoors tasks;
2. reflective journaling on related texts and thematic ideas; and
3. sharing in group discussions.

This is a basic model that works across the volume, but each essay offers different types of assignments that can accompany any trip, from storytelling to eco-soundscapes, journaling to field observations of the more-than-human world. We have pulled together here key lessons we've learned from our various experiences of bringing students out of the classroom and into the field and offer them here as a suggested framework for integrating outdoor experiential learning into any course.

Define the Field Loosely

While all contributors here center place-based and experiential learning in their ecopedagogies, they imagine "the field" in wildly different ways. We have found that getting students out the door is the essential component of our courses, and there are so many possibilities as to what awaits them on the other side. Examples of the field from our collection span from factories in urban France to the back-country of Alaska; campuses in Montréal and suburban New Jersey to canyons in Utah; ponds that are a stone's throw from the university to a virtual Walden Pond; and zoos to zoo cams. When we expand our sense of what "the field" might encompass, we remove some of the barriers to experiential learning that we have perhaps erected ourselves. We work within the environments that both we and our students can access, and we design courses, assignments, and activities that, using high-impact practices (HIPs), engage them intimately with these places.

Bear in mind that your conception of the field, the natural world, or wilderness will largely differ from your students'. Being conscious of your own biases about what constitutes the field or "nature" can allow you to take a step back and recognize opportunities that will energize your students even if they aren't venturing deep into the wild. Seeing through your students' eyes can transform your pedagogy in many different contexts. Ellen learned this on a day hike with an "Introduction to Literature and Environment" class. She fretted as the trail she'd chosen had been paved since her last visit, and an unexpected snowstorm obstructed the mountain views she'd been promising students. Apologizing all over herself for this absolute failure of a hike, students shook their heads and spoke up: "I didn't really notice the trail was paved until you kept mentioning it. I'm just happy to be out here, especially since I've never been before. The snowstorm makes it feel more like wilderness!" When we adjust our definition of the field and meet students where they're at, it can open new paths for all of us to embody places and engage with the more-than-human world in exciting and meaningful ways that we hadn't expected.

Make the Field Mandatory

Sending students out of the classroom is the central tenet to all the approaches collected here. As we outline above, widening the lens to encompass a range of fields is an important first step to making experiential learning accessible to all your students. We have discovered three key strategies that can help further facilitate the process.

Make it an Assignment

As she described in the opening of this book, Ellen learned, as an undergraduate, to trick herself into making time for a weekly outdoor assignment. Like her younger

FIGURE 1.1 A student is proud to have reached Panorama Point, at Mount Rainier National Park, during a snowstorm.
Source: Alex Hau Nguyen.

self, your students are probably spread thin and might question whether sitting by a creek for an hour each week is a valid use of their study time. *How will this help me in my career? I don't have time for this on top of homework and my job. Is this even going to be on the exam?* We understand why students might express skepticism toward college work that looks vastly different from anything they've encountered in other courses, particularly in the arts and humanities. They might not recognize something that seems a lot more like play and fun as a worthy academic pursuit. To counter this, it's important to frame experiential learning as homework. Doing so helps students justify taking time for an outdoor excursion. Ellen discusses this at length later in her essay, but we offer here this seemingly simple tip to help you combat what will likely be an initial assumption your students make. When we frame experiential learning as an assignment, as part of the course curriculum, as something that is assessed and graded—even if only in terms of being complete/incomplete—we signal to students, in a way they have been taught to understand, that this assignment has stakes. Even if the stakes are low, having them in place creates an understanding that this is an expectation for completing, and succeeding in, the course.

We can't stop there and simply say, "It's homework, and it's good for you. Just trust us." It's incumbent upon instructors to help students understand the value of experiential learning. Some of us in this collection use reflective writing as a means for inviting students to examine their experience and explore what they take away from it, while others use class discussions to create a space for students to consider what they gained from getting out, both on an individual level and in a broader sense of how it can impact their relationships with wider communities, both human and beyond. We must equip students with the knowledge, skills, and language that will allow them to appreciate that just because it's not on the exam doesn't mean they aren't growing as scholars, and humans, through this experience.

Introduce it Early

Chances are, when students enroll in your arts, humanities, or even social sciences course, they aren't expecting to spend time outside of the classroom. While faculty across the globe are doing good work to build experiential, outdoor learning into these disciplines, it's still not the norm, and we need to let students know right off that our courses might look a bit different from the literature courses, for example, they've taken in the past. It's also important for us to recognize that students come from a range of life situations and circumstances, and that these can be in flux even during the term they spend with us. While some may have the luxury of focusing all their energy on their studies, it's unlikely that this is true for many contemporary college students. Giving them time to plan for accommodating work schedules, childcare, and other responsibilities is essential to making field excursions accessible to all our students. Even if you simply plan to stroll around campus during class, students appreciate a heads-up so that they can dress for the weather and being on their feet longer than usual. Longer forays require even more advance notice. Whether they'll need to spend an hour outdoors on their own time occasionally throughout the term or participate in a weekend class trip, informing students from the first day that the course includes out-of-class learning experiences—and providing the dates, details, and expectations—can lead to higher participation in these outings.

Give Students Options

There will be students who, for various reasons, cannot participate in a trip to the field that occurs outside of class time. This might be due to a lack of time, resources, schedule conflicts, or personal matters that they may not feel comfortable sharing. Furthermore, students arrive in our courses with different levels of comfort, skill, and ability; it's our responsibility as instructors to develop experiences and excursions that can meet this range of needs. Giving students options for how they get into the field can ensure that they all have an opportunity to learn from their experience, even if it differs from their classmates'. Several contributors in this volume outline alternatives they offer students, recognizing that a one-size-fits-all approach is neither equitable nor practical. For example, Judson Finley and Keri Holt taught an honors class where many engineering students take a critical professional examination over Fall break when the trip was scheduled. While their three-day class excursion was required, they provided students with viable options to experience local public lands as an alternative.

Recognize the Limitations

Integrating field excursions into your course isn't necessarily easy. It will require additional work, and you'll have to navigate some challenges. Many disciplines, especially in the arts and humanities, lack a "field culture" and, as a result, don't have the experience and infrastructure for organizing trips. The first time Ellen inquired into using university vans to transport her literature students to a local park, the administration looked at her like she was demanding an all-expenses-paid trip to the moon: "We're just not set up for humanities courses to do that. That's something they only do in the sciences." Depending on your institution, you might face a maze of red tape and baffled administrators who don't know what to do with you. Persistence can pay off, though, and talking with colleagues in disciplines that regularly take students into the field can offer you some guidance on how to make outdoor, experiential learning part of your course. The contributors in this volume give us hope that we can continue to push the boundaries and expectations of which courses take students out into the field.

The challenges don't end with university culture. Many students will also lack the resources, time, or know-how to get into the field. One particularly crucial piece of this is gear. Gear is important, and not everyone has it. Our contributors offer a variety of ways to solve this problem. Some schools have outdoor recreation programs, which might also provide low-cost rentals to students. In some areas, public outdoor programs might also have gear lending libraries. Inevitably, some of your students will be outdoor enthusiasts who can share gear with classmates, and it's worth scanning your own gear closet (and those of your outdoorsy friends) for any surplus gear you can loan or donate.

Access is another limiting factor. Not everyone has a vehicle, and you may find that many students rely on public transportation. Furthermore, not all universities have easy access to the field beyond campus. Asking students to meet on campus and then carpool is one possible solution. Choosing a location served by public transportation is another approach. You might also mosey over to the offices of your colleagues in the sciences and explore the possibility of borrowing vans from their fleet. Some schools have partnerships with vehicle rental companies that offer discounted rates for class trips. This, of course, comes with additional costs, but it could be a more feasible way to take students further afield.

To address concerns of costs, our contributors take several approaches. Course fees are the most straightforward method for covering the financial costs of field excursions, and it's a common practice in disciplines with established field cultures. Since course fees are built into students' tuition, they can use financial aid to pay them. The drawback is that high course fees might be prohibitive for some students, especially those who rightly worry about becoming saddled with student loan debt. We've had success in applying for grants, both internal to our institutions and on a federal level through the National Park Service. While grant writing takes additional time, doing so can establish more sustainable and equitable access to the field for our students. The NEH's Humanities Connections Grants generously support the development of interdisciplinary courses that center experiential learning and could be a great way to forge connections with your colleagues across disciplines and establish exciting and important new curriculum for your school. As our contributors ultimately demonstrate, though, you need not venture far to provide students an impactful field experience, and this volume highlights many low- and no-cost options for getting students outdoors.

Technology is friend and foe. Composition pads are perfectly functional, but smart phones, tablets, and other equipment can change the experience. While you can do a lot with a cell phone, not everyone has one. As you design outdoor activities and assignments, be conscious of the assumptions you may make about access to technology, and offer alternatives so that everyone can participate. Reaching out to staff in your institution's IT office is a good place to start, as is the library or other offices on campus that support student learning. In many cases, they may have equipment that students can check out and use free of charge. Partnering or grouping up students is another potential way to address the gap in access. Technology can also open up new ways of engaging with the field, and teaching during the COVID-19 pandemic has pushed us to consider how we can modify our courses to function in a remote context. Some contributors here describe employing zoo cams and National Parks online tours as ways to allow students to participate in field observations from their own homes. Such programs can work to create more accessible "outdoor" spaces for observation and reflection, both during these extraordinary times and beyond.

We will have to do some extra work to get our students out into the field, but once we build a framework and gain experience, it will get easier. Over time, we can perhaps change the institutional culture and broaden our sense of who goes out into the field, and why.

Cultivate Diversity, Equity, and Inclusion

There is a disproportionate access to the outdoors that has its source in economics, geography, historical trauma, and settler-colonial relations. It's simply not enough to *invite* students from marginalized groups to the field—instructors need to put in the work to *bring* them there if we're going to cultivate a culture of diversity, equity, and inclusion in outdoor experiential learning. Addressing the common barriers that keep students out of the field—resources, time, access—is an initial step we can take. Here, we'll point to three additional areas to consider; many of the contributors in this volume also offer their own suggestions for fostering DEI and developing anti-racist ecopedagogies.

Recognize Student Concerns

Not all students will view the natural world as a peaceful place for quiet contemplation and exciting adventure. There is a long history of physical violence against marginalized groups in rural and wild places, and students carry the weight of that genetic and historical trauma with them. Students have learned through family stories, courses, and various media, that the outdoors can be a site of violence and harm to people who look like them. As instructors, we must recognize this as a valid concern, acknowledge and discuss it in our courses, and work to mitigate that fear for our students. Assigning texts that confront this reality is one avenue for creating space for such discussions, as is incorporating materials that illustrate the wide array of people who go into the field. Offering students a wider representation of voices and faces in course texts and materials signals to them that people with whom they can identify belong outdoors.

Decolonize Outdoor Education

We, along with other non-Indigenous scholars in this collection, employ the term *decolonize* with some caution, understanding that there are broader debates about this term and its use to characterize pedagogies without acknowledging the potential for perpetuating settler worldviews. Efforts to decolonize ecopedagogies within the context of this collection seek to engage BIPOC communities, non-human animals, and more-than-human places on equal footing. As Mary A. Jackson explains in Chapter 12 of this volume, "Vital to reconfiguring dominant narratives and subversive practices is the inclusion of decolonizing methodologies and traditional ecological knowledge (TEK)." Contributors to this collection are less concerned about the production of knowledge than they are its dissemination.

Louie et al. (2017, p. 21) identify 25 principles for embedding Indigenous ways of knowing into transformative pedagogies that parallel Smith's (2012) foundational decolonizing methods. These principles run throughout the collected experiences and lesson plans. In Chapter 7 of this volume, Kevin Maier explores this academic discussion in more depth, noting that his "tentative efforts to decolonize [his] pedagogy hinge on course outings that show students that sometimes the act of going outside to experience the environments of southeast Alaska in ways we've been told are not only innocuous and fun, but educational and good, can in fact be detrimental to Indigenous self-determination." Other contributors use strategic intent to develop ecopedagogies that set decolonizing as a primary objective. Decolonizing outdoor education is a goal that is best served by approaching it collectively, by sharing a range of different stories and experiences, and by committing to meaningful actions that go beyond passive approaches to inclusion. To this end, we intend the diverse, communal structure of this collection to foster continued conversations about decolonizing outdoor education.

Meet Students Where They're At

As Kelly Branam Macauley et al. illustrate in Chapter 10 of this collection, sometimes getting students out into the field might be contingent on bringing others from outside the class as well. In their experience, creating a truly inclusive field school meant inviting students to bring along family members and finding funding and gear to accommodate them. Without doing so, they would not have been able to engage the Indigenous students they'd hoped to. In other instances, they also had to build into their grants "caretaker stipends" in order to facilitate bringing some invited guests who needed help with transportation to the field site. We've come to learn the importance of this in more local excursions as well. After a student who was also a Somali refugee asked Ellen if she could bring a friend with her on a class trip to a local park, she understood that similar concerns about venturing out beyond campus with a group of relative strangers might be uncomfortable for others. Since then, inviting students to bring along a friend, family member, kids, or even the dog has opened opportunities for students to share these experiences with those they love while also accommodating students who otherwise might not be able to attend, whether it's due to personal safety concerns or lack of childcare.

Create Community, Not Competition

One of the major outcomes of field learning is that it breaks down competition between students in favor of collaboration. When we shift the focus to their experience, students have both a common source from which to draw as well as individual responses and perspectives to express. Whether it's doing camp chores, building an emergency fire, observing a heron, or recording sounds as a team, having shared goals and learning outcomes not linked to exam content can foster a sense of community within a course. Participating in the activity is the point, not proving who is the smartest student. When we take students outside their comfort zone, they can lean on each other in this new environment. As several contributors in this collection demonstrate, a mix of students with outdoor experience and those without is a valuable blend. Those with more experience can lend support to those who may be in the field for the first time, and this can also decenter instructors and reinforce the value of students learning from each other. Privileging discovery, observation, and reflection can energize students, validate their ideas, and build a sense of community in the field, which will follow you back to the classroom.

Keep it Simple

This volume represents a wide range of excursions from low investment to high investment. We've found that working within our means, our resources, and our comfort levels has been important to building experiential learning successfully and sustainably into our courses. Bear in mind that a three-hour tour can be just as impactful as a three-day camping trip. Starting small can help to get your foot out the door without overwhelming yourself, especially when you're working within a discipline or school that doesn't have an established field culture. As many of the chapters in this volume show, there is much you can do right outside the classroom, even on an urban campus. Over time, you'll be able to build experience, make connections, identify and secure resources, and eventually organize more extensive trips, if you'd like. Ellen started by taking students around her urban campus, then moved up to local parks. After a few years, she found ways to get students to a nearby National Park for a day trip. She's now putting the finishing touches on a team-taught, interdisciplinary course—developed thanks to a grant intended to support such courses—that will get students out into the field for 2–3 weeks during the summer term. The smaller, simple trips are impactful in their own right, but they can also serve as a great entry point for building toward more intensive excursions.

Keep it Safe

Having well-defined travel plans and safety protocols eases the minds of students, families, and university administrators alike. Successful and safe excursions ensure the continuity of programs and reinforce that students will return into the field, which should be a primary outcome of field-based learning. Some schools have well-established waivers and forms, but you might also need to tailor something specific to your field excursion. Working with your institution's risk management staff to develop risk waivers and medical release forms can offer helpful guidance in navigating these new waters.[1]

Teachers embarking with students on trips that venture further or longer should acquire the skills necessary to ensuring everyone comes home safely. We highly recommend Wilderness First Responder (WFR) training, or at the very least Wilderness First Aid (WFA) training, and critical first aid and safety equipment. Many organizations

FIGURE 1.2 Students participating in a sensory experience activity at the University of Washington Tacoma's campus Giving Garden.
Source: Ellen Bayer.

now offer hybrid WFR models, in which you conduct much of the course through self-paced online modules and then meet in-person for five days of intensive hands-on training. We were able to use professional development funds to cover our tuition costs in these programs. Using university funding to reinforce student safety is not a hard sell.

Partnerships can help, too. Team teaching, as several of our contributors illustrate, is an effective solution that bridges experiences and makes successful day, overnight, and extended trips. Shared leadership can make trip planning, logistics, budgeting, risk management, and outdoor education and design more accessible and less intimidating, especially to teachers who are new to bringing students to the field.

Start Change Now

Generational shifts are slow to take hold. While teachers across the disciplines are embracing HIPs in their classrooms, outdoor experiential learning is not yet commonplace across the arts and humanities. As educators, we have the opportunity to be active participants in loosening the grip of the traditional classroom and empowering our students to learn in ways that allow them to recognize themselves as stakeholders in their local and global communities. We understand the hesitancy some teachers may feel toward bucking tradition and taking what they might feel to be pedagogical risks, especially when growing numbers of faculty find

FIGURE 1.3 Students paddling the intersection of built and natural environments in Tacoma, Washington.
Source: Ellen Bayer.

themselves in unprotected positions. We believe this volume illustrates the academic and personal rewards, for students and faculty alike, that outdoor experiential learning fosters, showing that we can still achieve our universities' expectations for rigor and specific learning outcomes. Our contributor Jeremy Chow speaks to the importance of being made uncomfortable, noting that, "in that discomfort rises pedagogical innovation and philosophical reawakenings." We agree, and we hope the essays collected here will offer you the encouragement and inspiration needed to begin, or deepen, your own journey into the field.

Note

1 See our companion website for sample travel plans and itineraries, risk waivers, and medical release forms.

References

Basso, K. H., 1996. *Wisdom sits in places: Landscape and language among the Western Apache*. Albuquerque, NM: University of New Mexico Press.

Bowers, C. A., 2008. Why a critical pedagogy of place is an oxymoron. *Environmental Education Research*, 14 (3), 325–335.

Dillard, A., 1974. *Pilgrim at Tinker Creek*. New York: Harper.

Gruenewald, D. A., 2003. The best of both worlds: A critical pedagogy of place. *Educational Researcher*, 32 (4), 3–12.

Kahn, R., 2008. From education for sustainable development to ecopedagogy: Sustaining capitalism or sustaining life? *Green Theory & Praxis: The Journal of Ecopedagogy*, 4 (1), 1–14.

Kimmerer, R. W., 2017. Learning the grammar of animacy. *Anthropology of Consciousness*, 28 (2), 128–134.

Kolb, A. Y., and Kolb, D. A., 2017. Experiential learning theory as a guide for experiential educators in higher education. *Experiential Learning and Teaching in Higher Education*, 1 (1), 7–44.

Lewis, L. H., and Williams, C. J., 1994. Experiential learning: Past and present. *New Directions for Adult and Continuing Education*, 62 (Summer), 5–16.

Louie, W. L., *et al.*, 2017. Applying Indigenizing principles of decolonizing methodologies in university classrooms. *Canadian Journal of Higher Education*, 47 (3), 16–33.

Machlis, G. E., and Jarvis, J. B., 2018. *The future of conservation in America: A chart for rough water*. Chicago, IL: University of Chicago Press.

Smith, L. T., 2012. *Decolonizing methodologies: Research and Indigenous peoples*. New York: Zed Books.

TallBear, K., 2015. An Indigenous reflection on working beyond the human/not human. *GLQ: A Journal of Lesbian and Gay Studies*, 21 (2–3), 230–235.

2

COMPOSING WITH INFRASTRUCTURES

Parapersonal Pedagogies for Environmental Humanities Classrooms

Andrew Niess and Davy Knittle

In this chapter, we offer two undergraduate-level assignments in the environmental humanities. One is a triple timescale composition that invites students to record and interrogate how they relate to infrastructures that subtend human and more-than-human life at multiple scales. The other is an infrastructural autoethnography assignment that asks students to spend one day documenting their relationships with built and nonhuman environments and the labor that sustains them, and then to reflect on their observations. Because of the range and adaptability of their techniques—sound recording, composition, juxtaposition, writing, auto-ethnography—these assignments will be useful for a wide range of environmental humanities courses. While they draw on our different disciplinary training in English, urban studies, music, and sound studies, both assignments are occasions for students and teachers to develop relationships with time, to think with multiple senses, and to attend to interlocking scales of analysis. These assignments adapt for the classroom the work of scholars in Black studies, Indigenous studies, queer and trans theory, performance studies, and the environmental humanities who argue that developing new ways of thinking and being "in good relation"[1] is crucial to understanding and disrupting patterns of racialized environmental harm and dispossession. We believe, with others, that the destructiveness of contemporary capitalist systems stems partly from settler relationships with time, text, and sexualities that are constitutive of Western modernity and its cis-heteronormative ideas of progress (Latour 1993; Freeman 2010; Mignolo 2011; Rifkin 2017; Vázquez 2017; Bjornerud 2018; Keeling 2019; TallBear 2019). Because settler ways of relating are based fundamentally on possession (Moreton-Robinson 2015), and because they pervade the present and poison the imagination of futures (TallBear 2019), we offer these assignments for environmental humanities classrooms as responses to the need to develop, practice, and model ways of thinking, creating, teaching, and living "in good relation."

DOI: 10.4324/9781003221807-2

We use these assignments to ask how humanities courses might encourage students to think beyond the human and to think about human lives as closely entangled with more-than-human lives as practices made urgent by the exigencies of the climate emergency. We follow, for instance, Zakiyyah Iman Jackson's (2020, p. 28) "questioning of the universal liberal human project." Jackson is interested in how "the human" as a category has overprivileged the needs and desires of some humans while simultaneously weaponizing the category to harm others. We link Jackson's critique of the human as a racialized category used to justify violence against some humans with critiques in the environmental humanities that object to the use of "the human" as a category that has also condoned harm against more-than-human beings and environments (Wynter 2003, TallBear 2015). We encourage students to think beyond a universalizing relationship to the human to address systems of subjection that harm both human and more-than-human life. By challenging students to question how they are part of and how they interact with more-than-human beings and environments, these assignments give students other options for negotiating and resisting "the realm of the human, a colonial realm within which the human functions as a sovereign power" (Halberstam 2020, p. 7).

These assignments echo questions central to counter-hegemonic scholarship. For instance, to position ourselves within complex scales, "how can we think in times of urgencies *without* the self-indulgent and self-fulfilling myths of apocalypse, when every fiber of our being is interlaced, even complicit, in the webs of processes that must somehow be engaged and repatterned?" (Haraway 2016, p. 35; original emphasis).

To orient ourselves toward time, "how do we produce a vision that enables us to see beyond our immediate ordeals?" (Kelley 2002, p. x). In designing these assignments, we approach such questions by encouraging students to use multisensory attention to track their relationships with infrastructure.

For the purposes of this essay and these assignments, we consider infrastructure as that which "has determined the flow of resources and risks" (Hurley 2020, p. 3). Infrastructure "binds us to the world in movement and keeps the world practically bound to itself" (Berlant 2016, p. 11). Infrastructure is composed of multispecies beings, the built environment, and the management of environmental resources, and it is through observation and composition that these assignments invite students to become attuned to the physical and ideological systems that circumscribe the frameworks of perception. How do marginalizing ideologies subtend infrastructures just as infrastructures subtend our everyday interactions with structures, beings, and systems? These assignments invite students to practice being attentive to how their own beliefs and ideas are complexly entangled with the infrastructures that undergird their daily lives.

To help students practice attending to the systems that undergird relations between human and more-than-human life where they live, the assignments invite students to cultivate three skills: (1) *perceiving* how layers of human and more-than-human lives compose the conditions of their own daily spaces and practices; (2) *composing* with sound, image, and text to communicate such perception interpersonally; and (3) *transforming* the quality of relationships with human and more-than-human life. These skills are designed to effect transformation at the relational scales of the intrapersonal (how one perceives one's own orientation and habits of mind), interpersonal (how one interacts and communicates with others), and parapersonal (how one relates to human and more-than-human life). A primary goal of teaching students to think across scales is to bridge the implicit distance that many students associate with the category of "the environment" as a scale apart from their everyday lives. We believe that the cultivation of these skills and an awareness of these interlocking relational scales is integral to resistant thinking and practice. To counter universal humanism and its inherent racialized violence, we need to dedicate our pedagogies to the cultivation of "a new epistemology and transformative approach to *being* … rather than the extension of human recognition under the state's normative conception" (Jackson 2020, p. 28, emphasis in original). We hope that the cultivation of these skills through three related design principles—relating differently with time, multisensory thinking about and with infrastructure, and attention to interlocking scales of analysis—may play a small part in encouraging students to direct their energies toward such a transformative approach. To this end, we delineate this set of design principles shared by two assignments: a triple timescale composition and an infrastructural autoethnography. We begin with a description of those design principles and then share the details of our assignments. We conclude with further considerations for future assignment design and collaboration among environmental humanities educators.

Design Principles

Transforming Relationships with Time

Our assignments help students develop ways of attending to environmental change and relationships with the more-than-human. To cultivate such attention requires explicitly addressing the timescales of more-than-human lives and environments. Environmental change is often narrated as consigned to an ominous near future or as suddenly crashing into the present. Such thinking inures us to slower processes that are "neither spectacular nor instantaneous, but rather incremental and accretive" (Nixon 2011, p. 2). Such forms of slow violence might look and sound like the gradual increase of annual rainfall, rise in monthly average temperature, or deteriorating air quality.[2] Slow violence may also unfold as projects by which humans more directly cause gradual and accumulative harm, such as university-backed gentrification that displaces longtime residents over decades (Lubrano and Gammage 2019; Richardson, Mitchell, and Franco 2019).

We propose methods for attending to processes "not viewed as violence" because their forms of violence are neither sufficiently spectacular nor speedy (Nixon 2011, p. 2). The goal of developing such methods is to attune students both to their relationships with infrastructure and more-than-human environments and to think about how such a change in attention is related to how they conceptualize justice. How these assignments address our students' practices of attention is closely related to broader efforts to mobilize our universities' wealth and resources "to contribute to the politicization of presently de-politicized everyday spaces" (Smucker 2017, p. 38). Too often, deploying novel methods and assignments in the classroom stops short of this goal by merely constructing "a special sphere that houses our socially enlightened identities (and delusions)" but which fails to mobilize such knowledge to realize political transformation (ibid.).

Political transformation can occur within and across intrapersonal, interpersonal, and parapersonal levels. Such political change follows from a parapersonal sensibility that attends to human–nonhuman relations existing alongside but not separate from one's own life. With this relational awareness, one exists neither as the cause nor the effect of other life. Instead, a parapersonal view is oriented by the quality of existence with and alongside other life. We seek a pedagogical reorientation toward infrastructure, to listening at parapersonal scales and composing with non-linear temporalities. In so doing, learners practice empathizing not only with their own rhythms but also with the rhythms of the life around them.

Multisensory Thinking

Multisensory thinking refers to attentional practices oriented toward meaning- and sense-making that exceed human language. Our assignments give students the opportunity to practice perceiving, arranging, and sharing information that takes extralinguistic forms. We are interested in multisensory thinking because it helps students become aware of the more-than-human communities and infrastructures in which they participate, and it empowers students to mobilize that expanded attention to reframe the ways of knowing that they value.

As we use our assignments to encourage students to develop more expansive practices of sensory attention, we follow anthropologist Eduardo Kohn's assertion that "we conflate representation with language in the sense that we tend to think of how representation works in terms of our assumptions about how human language works" (Kohn 2013, p. 8). Kohn argues for an attention to more-than-human methods of communication, which he argues are distinct from but importantly related to human language. As he suggests, "it is through our partially shared semiotic propensities that multispecies relations are possible, and also analytically comprehensible" (ibid., p. 9). Sonic creativity and attention to the production of difference between rhythms, synchronicities, and harmonies plays an important role in multisensory meaning-making.[3] We hope our assignments enable students not only to approach and comprehend relations that exceed the human but also to recognize the political and environmental value of those relations.

To this end, multisensory thinking helps teachers and learners to turn away from the colonial tendency to reduce world to text and expunge non- or extra-linguistic ways of knowing and making life. Curricula that uphold "textual fundamentalism" (Conquergood 2002, p. 147) tend to invite learners to passively receive decontextualized knowledge while arranging content knowledge within narratives that value individualism and unlimited linear progress. Pedagogies that "fail to imagine alternatives to textual supremacy—or as Édouard Glissant calls it 'The world as book, the Book as world' [Glissant 1997 [1990], p. 25]—are the enduring remnants of academic institutions' constitutive coloniality and extractivism" (Niess 2021, p. 4). In our assignments, we think beyond the conflation of representation and human language, as we encourage students to link their practices of sensory attention to their awareness of the multispecies communities and infrastructural contexts in which they participate, such that they can consider the political stakes of attending to the parapersonal.

Attention to Interlocking Scales of Analysis

We approach infrastructure as fundamentally multiscalar. In our thinking about scale, we follow arguments in the environmental humanities that center the practice of thinking across scales as necessary for describing uneven manifestations of environmental harm. For environmental humanists, ecologists, and geographers, ecological scale "consists of both temporal and spatial dimensions" and "there is no single correct scale for ecological research" (Sze 2016, p. 178). In both assignments, we ask students to practice becoming attuned to different scales of analysis. Students learn such attunement by noticing the scales within which they are accustomed to thinking, and how those scales rely upon others. For example, a student recording the soundscape of their university library might connect their own typing or turning the pages of a book to the hum of the building's heating system. They might then wonder about how the energy powering the heating system was sourced, and thus be drawn from their own observations of their immediate surroundings to a larger web of resources on a regional, national, international, or planetary scale.

We also ask students to situate their own bodies and lives in different relations to scale. For instance, as they focus on a sonic environment that may include a house sparrow, an airplane flying overhead, a car door slamming, and water running in another room, we ask students to connect the scale of those individual actors to the larger scale of the municipal water system to which their home is connected or the ecoregion in which the sparrow lives. We draw upon anthropologist Gabrielle Hecht's assertion that "scale is not just about size or granularity. It is also about categories: what they reveal or hide, the ways in which they do (or do not) nest. And it is about orientation: how we position ourselves, what we position ourselves against, and what comparisons such locations do (or do not) authorize" (Hecht 2018, p. 114). In our consideration of scale, we ask students to name the categories of analysis to which they are most immediately drawn. We ask them to use their observations to consider what their scales of analysis reveal or occlude. Given the opportunity to think of scale as a tool kit, we encourage students to reflect on how they might engage different scales to focus on specific resources or actors in the systems they analyze.

For example, in her infrastructural autoethnography, one student, Leah, drew on the methods of multiscalar analysis to explain how her access to and use of infrastructural systems was dictated by her position within a racial capitalist system. Leah noted:

> Like the vast majority of people in the United States, I am entirely dependent on infrastructure for every aspect of my life. In theory, I knew this to be true, but my privilege has afforded me the luxury of never having to think about it. If I am cold, I turn on the heater in my room; if I am thirsty, I fill up a glass with filtered water from the sink; if I am antsy, I go on a drive around town. My status as a wealthy, white, cis-gender woman means that I do not have to be conscious of these decisions. For the privileged people of the world, there exists an assumption that infrastructure will always be there to meet our needs. This is contrary to the many other communities in the US who feel as though infrastructure is a source of stress, anxiety and betrayal.
>
> *(Leah, student paper, University of Pennsylvania, 2021)*

In dialogue with Leah's reflection on the multiscalar implications of her own relationship with infrastructure, we hope that students will experiment with categories like "the environment," "the university," "the ecoregion," and "the city" that at once present a scale of analysis and indicate the necessity of a range of scales. We also hope that in interdisciplinary courses in the environmental humanities, students connect their examination of a range of scales to the affordances of the disciplines with which they are engaged. For instance, students might connect their interest in the psychological impact of living with a contaminated water supply as explored through Jennifer Haigh's Pennsylvania fracking novel *Heat and Light* with a national-scale sociological analysis of the impact of water contamination on mental health. An attention to infrastructure requires students to be intentional about the tools of analysis and formats of communication they are inclined to use and how the scales they choose form their particular orientation to their surroundings.

Relational Outcomes

Such design principles—transforming relationships with time, multisensory thinking, and attention to interlocking scales of analysis—are common to assignments across the environmental humanities and help students to situate their experiences even when they may "believe that they have no perspective, that their discourse is not conducted within very narrow boundaries" (Cordova 2007, p. 158). These design principles represent one response to the "[call] for an effective response from academics, scholars, [and] students … to examine how we come to produce, interrogate, validate, and disseminate knowledge" (Dei and Jajj 2018, p. 2). We hope that a consideration of these design principles aids in producing pedagogical conditions for relational outcomes that are more empathetic, embodied, and attentive to human and more-than-human scales of life.

We have focused so far on delineating a set of principles for designing environmental humanities pedagogy. What follows are descriptions of two assignments that model these principles. While we offer these specific assignments, we believe that these principles are sufficiently general, and we encourage you to adapt them for your own pedagogical needs.

Assignments

Triple Timescale Composition

In this assignment, students record and compose with sound to connect with their surroundings as conditions for, rather than parenthetical to, their living and learning. Using sound recorders and audio-editing software, students compose or put into sonic relation three different timescales—namely, their own "meso" quotidian rhythms with one "micro" and one "macro" timescale. Using "sound as a critical mode of relating … across species and materialities" (Feld 2015, n.p.), students contend with the aesthetic problem of connecting the disconnected through practices of attunement to multiple timescales. In juxtaposing micro, meso, and macro timescales and sharing their compositions with others, students are challenged to change how they perceive (intrapersonal), how they communicate their perceptions (interpersonal), and how they relate to infrastructures that subtend human and more-than-human life (parapersonal). To avoid an overly prescriptive, and thus less useful, assignment, the sections below provide key considerations to aid the widest range of educators in adapting the assignment to their own pedagogical, institutional, and community needs. To devise your own triple timescale composition assignment, consider the following ideas on equipment, assignment outcomes, adaptations for multiple learners, definitions, activities for embodied learning outside of the classroom, and models of Andrew's own audiovisual projects to share with students.

Equipment

The following equipment is required:

- a sound recorder (e.g., phone, field recorder with SD card); and
- a computer with audio-editing software (e.g., Audacity, Garageband).

To make cost-effective sound recordings, students may use smartphones with apps like Voice Memos or Voice Recorder. If students do not have access to smartphones, consider connecting with your institution's music or technology departments to inquire about access to field or voice recorders, which can cost as little as US$20. To create their compositions, students use audio-editing software such as Audacity (free, open source, cross-platform) or GarageBand (free, for Mac only). With this software, students import audio files; create three or more tracks; move, trim, copy, and paste audio clips; and export projects as audio files.[4] Consider inviting librarians, music technologists, or digital literacy experts at or outside of your institution to your course to assist you and your students.

Outcomes

By completing this assignment, students will be able to:

- Make sound recordings that attend to infrastructure and micro, meso, and macro scales.
- Create a composition that juxtaposes these recordings.
- Communicate changes in their own perceptions of timescales and how they relate to infrastructure.

In adapting this assignment, teachers are encouraged to:

- Establish a public-facing repository for sharing student projects with surrounding communities as a contribution to free, public knowledge.
- Cultivate relationships with and invite into your class community members who are differently positioned toward your institution: neighbors displaced by university development; nearby Indigenous communities; community science practitioners; museum curators; ecologists, botanists, and wildlife conservationists. Foster reciprocity in these relationships.
- Frame assignments such that they may heal and nourish you, your students, and your community and not harm or exhaust.

Adaptations

Multisensory thinking need not be limited to the domain of audibility. Cultivating different relationships with time and attending to interlocking scales of analysis may be adapted for the range of deaf and hard of hearing learners. Encourage students to juxtapose their daily rhythms with other scales by attending to them in ways that make the most sense for each learner's body. For instance, invite them to juxtapose images, drawings, texts, or videos that communicate the intensities they perceive at multiple scales.[5]

Activities and Models

Introduction to Field Recording

Consider initiating a brief sound walk, listening back to students' recordings together, and talking about what they are hearing. With sound recorders, gather students outdoors in a circle. If using field recorders, encourage students to wear headphones so they start listening through the recorders. After ensuring that everyone is ready to record, ask students to face outward and walk or roll in a straight line for five to ten minutes while recording. Upon returning, facilitate a conversation about what students heard. This is an opportunity to listen for whether and how students are connecting what they hear to infrastructures, multiple scales of time and space, and broader concerns guiding the course.

Define Micro, Meso, and Macro Timescales

Because semester timescales do not map onto slower timescales of change, for this assignment's purposes we define "micro" as unfolding at a scale that is less than one day, "meso" as unfolding across one day, and "macro" as unfolding at a scale that exceeds one day. Recordings of micro timescales might represent the number of times waste is disposed of at a busy trash receptacle for twenty minutes. Recordings of meso timescales might represent each time a student rides public transportation during the day. Recording macro timescales will likely be most challenging because they exceed what can be observed in a single day. Below are ideas for attending to macro timescales.

Interviews with Elders

Students interested in documenting, for instance, gentrification will have different first-hand access to how a neighborhood has changed over decades than will elders who have lived there much longer. Students may develop relationships with such elders, request to record conversations with them, and juxtapose their statements with micro and meso timescales. Before approving such projects, review with students the ethics of conducting and recording interviews, including obtaining narrators' informed consent.[6] Students may also consider mobilizing archival audio of news reports to similar effect.

Satellite Images

Encourage students composing with images to consult satellite imagery to visualize how land has changed over time. Students may use Google Earth Pro or the US Geological Survey's EarthExplorer.[7] For a model of stitching together satellite imagery with sound recordings, see Andrew's film *Particulate Matters*.[8]

Parapersonal Affinities

If students struggle to identify relationships between the three timescales, encourage them to explore similarities between human capacities that they observe in and around themselves and infrastructural capacities they observe at larger scales. For instance, students might interrogate the affinities between human breathing and steam vents; between eating and waste management; between drinking, drains, and bodies of water. Consider sharing Andrew's *a voice like* and *Pneumatography I* as audiovisual models that explore parapersonal affinities.[9]

Topics and Sites of Concern

If students struggle to put into relation their recordings of multiple timescales, encourage them to focus their recordings and composition on either a specific topic or site. Topics might include life's resilience despite obstacles or changes brought about by land dispossession or gentrification. Sites might include nearby superfund sites or bodies of water. Once students have determined a topic or site, it will be easier to bring to bear an attention to interlocking scales and thus to compose with what they have recorded and perceived.

Infrastructural Autoethnography

The infrastructural autoethnography assignment begins by introducing students to the concepts of ethnography, autoethnography, and infrastructure. Drawing on Deborah Reed-Danahay's (2017, n.p.) definition, the assignment defines autoethnography as "a genre that places the self of the researcher and/or the narrator within a social context … it reflects a view of ethnography as both a reflexive and a collaborative enterprise." Students begin the assignment by keeping an infrastructure journal for one full day in which they note the infrastructural systems (water treatment, heating, roads, transit, food distribution, telecommunications, etc.) that they use. After students conduct their observations, they write a narrative description of what they noticed about their relationship to infrastructure. The guidelines ask students to think about the systems that support their daily lives, the other beings that use those systems, and the labor required to maintain and operate those systems. While the practice of thinking about infrastructure may be new to students, the assignment is premised on the idea that interacting with infrastructure is a form of expertise that students share.

Although this assignment was designed specifically for the Spring 2021 urban studies course "Gender, Sex, and Urban Life" at the University of Pennsylvania, the assignment is widely adaptable. In environmental literature courses, students might compare their observations of infrastructure with infrastructural systems as they are represented in an assigned literary text. In interdisciplinary environmental humanities or environmental studies courses, students might focus on infrastructural systems that mediate local, regional, and global environments or which concretize global flows of resources. To introduce readers to the assignment, I focus on how students in this course approached the assignment and on how instructors might adapt the assignment for their courses.[10]

In their narrative descriptions, students' observations of infrastructural systems were inflected with how their own subject positions including race, gender, sexuality, and institutional privilege shaped the types of infrastructure they had access to. George,[11] who grew up in Houston, noted in his autoethnography that

> when a winter storm sent most of the state to historically low temperatures, the non-winterized state-specific power grid could not produce electricity to meet demand while it struggled to even remain afloat; this sent power offline for the majority of the state. Of course, the effects of the shutoff were not evenly felt; downtown areas remained lit up, creating a poignant symbol of the mostly empty office buildings lit up in the night as people froze in their homes.
>
> *(George, student paper, University of Pennsylvania, 2021)*

Another student, Natasha, wrote about how the racialization of infrastructural systems and their maintenance on Penn's campus shaped her emotional relationship to the campus as a Black student. Natasha reflected on the tensions of interacting with food systems. As a first-generation, low-income student whose meal plan is paid for by the university, she described being able to afford foods at Penn that her family could not afford at home. Her access to fresh food, she noted, changed dramatically once she adjusted to the options available to her on campus. Reflecting on eating a bowl of organic cereal and oat milk for breakfast, she noted:

> I couldn't help but feel like my parents would be disappointed in me at that moment. And it wasn't because I was doing anything wrong, but because I was doing everything right, and doing everything that I couldn't do when I was at home with my family.
>
> *(Natasha, student paper, University of Pennsylvania, 2021)*

In her reflection, Natasha described negotiating infrastructural privilege on campus as a tension between gratitude and shame. Sara, a white student, also noted having a conflicted relationship to stratified infrastructural systems. As she wrote of taking out her trash,

> This morning, I went to take out my garbage. I live on Delancey Street, notorious for a trash pile-up that accumulates when we go two, three, or even four weeks without garbage collection … Just a few moments after I set the garbage into the bins, I noticed … a private garbage collection service … I thought about the hundreds of West Philadelphian residents with trash pileups, for whom chartering private trash collectors is entirely outside of the realm of possibility.
>
> *(Sara, student paper, University of Pennsylvania, 2021)*

Students' reflections about infrastructure were shaped by their attention to racialized and gendered systems of power. Black intellectual historian Robin Kelley (2002, p. 19) argues that "the most radical art is not protest art but works that take us to another place, envision a different way of seeing, perhaps a different way of feeling." In the assignment, students used attention to scale, temporality, and infrastructure to see and feel differently, a goal common to many environmental humanities courses.

Conclusion

Both the triple timescale composition and infrastructural autoethnography assignments encourage students to contend with the problem of connecting the disconnected through practices of attunement. Students observe and recognize the systems that structure their daily lives. Students also experiment with changes to their behavior that affect what and how they notice in their immediate environments. Our hope is that the practices of attention that students learn in these assignments affect not only how they think about environmental questions and relationships with the more-than-human in their academic work, but also how they interact with the environments and infrastructures that support their daily lives beyond the classroom and that deserve care and respect.

One of the most pernicious problems troubling North American educational institutions is the maintenance and endurance of settler colonialism and its relational and knowledge practices. One transformational effect of attending to the quality of parapersonal relationships and of connecting micro, meso, and macro timescales is that it "can call non-Indigenous people (including those who do not fit well into the 'settler' category) to be more accountable to Indigenous lifeways long constituted in intimate relation with this place" (TallBear 2019, 38).[12] Both assignments offer students tools to bridge theoretical conversations about environmental justice and climate futures with actionable tactics for changing attention by giving students the opportunity to apply the theoretical methods of the environmental humanities to their relationship with their immediate environment and to conceptualize their physical context as part of a global system. We likewise encourage teachers to consider how their courses broadly, and these assignments specifically, can contribute to free, public knowledge by cultivating reciprocal relationships with community members. For instance, maintaining an online hub of students' compositions and autoethnographies is a way to mobilize university resources toward creating living archives of your communities and the changes they undergo. We hope that these assignments help students to practice applying theory to their own lives and, accordingly, expanding their practices of relation that extend from the relations inherent in turning on the faucet to the relations inherent in reading or listening to another's work. Providing students with the tools to connect these several kinds of relation prepares students and teachers to navigate the inevitable environmental ruptures that shape and will continue shaping their adult lives. These assignments contribute to this goal by encouraging students to practice attending to the human and nonhuman lives with whom they are in relation.

Acknowledgment

We offer this acknowledgment from a desire to work beyond what a land acknowledgment alone can accomplish to participate in collective efforts to actively return land to colonized peoples. At the time of writing this chapter, we were both affiliated with the University of Pennsylvania, a settler colonial institution whose late eighteenth-century formation followed from John and Thomas Penn's fraudulent seizure of land via the "Walking Purchase" of 1737, and which has contributed to the normalization of the displacement of Lenni-Lenape people in what is presently claimed as Philadelphia (see Barker 2018, esp. pp. 28–9).

For use by those who seek to honor Lenape people and territory, the Nanticoke Lenni-Lenape Tribal Nation offers the below land acknowledgment, which we have reused with slight modification. This land acknowledgment and more information about the Lenni-Lenape are available at https://nlltribe.com/land-acknowledgement.

The land upon which we wrote this chapter is part of the traditional territory of the Lenni-Lenape, called Lenapehoking. The Lenape People lived in harmony with one another upon this territory for thousands of years. During the colonial era and early federal period, many were removed west and north, but some also remain among the continuing historical tribal communities of the region: The Nanticoke Lenni-Lenape Tribal Nation; the Ramapough Lenape Nation; and the Powhatan Renape Nation, The Nanticoke of Millsboro Delaware, and the Lenape of Cheswold Delaware. We acknowledge the Lenni-Lenape as the original people of this land and their continuing relationship with their territory. In our acknowledgment of the continued presence of Lenape people in their homeland, we affirm the aspiration of the great Lenape Chief Tamanend, that there be harmony between the indigenous people of this land and the descendants of the immigrants to this land, "as long as the rivers and creeks flow, and the sun, moon, and stars shine."

Notes

1 Throughout her work, Kim TallBear foregrounds an Indigenous analytic of "being in good relation," which she poignantly states as "liv[ing] together in a good way here—as kin or as Peoples in alliance with reciprocal responsibilities to one another and to our other-than-human relatives with whose land, water, and animal bodies we are co-constituted" (TallBear 2019, p. 36). On sexualities and being in good relation, see TallBear (2018a, 2018b).
2 For a project that helps us attune to air quality changes, see Niess (2021).
3 On biosemiosis and musicking, see Tomlinson (2016).
4 For help with these skills in Audacity, see Audacity's online manual, especially "Tutorial—Editing an Existing Audio File," at https://manual.audacityteam.org/man/tutorial_editing_an_existing_file.html. For GarageBand, see the online manual, especially the pages "Import audio and MIDI files in GarageBand on Mac," "Edit audio regions in Audio Editor in GarageBand on Mac," and "Export songs to disk from GarageBand on Mac," at https://support.apple.com/guide/garageband/welcome/mac.
5 For examples that may inspire students, see the works of Deaf composer and sound artist Christine Sun Kim at http://christinesunkim.com/works/.
6 For a guide to ethics of recording interviews, see Oral History Association's "Statement on Ethics" at www.oralhistory.org/oha-statement-on-ethics/. IRB requirements for student coursework and interviews vary; check your institution's IRB policies on student research and interviews.
7 Google Earth Pro is a free desktop application available for download at www.google.com/earth/versions/#download-pro. For how to view a map over time, see https://support.google.com/earth/answer/148094?hl=en. The USGS EarthExplorer is available online at https://earthexplorer.usgs.gov/.
8 See https://vimeo.com/323348390, esp. from 5:54 to 6:42. This film is also available in Niess (2022).
9 Andrew Niess, "a voice like," https://andrewniess.com/a-voice-like, and *Pneumatography I*, https://vimeo.com/335996550. In "a voice like," I juxtapose sounds of my own breathing, eating, and waste with sounds of nearby waste management infrastructure.
10 Full assignment guidelines and five anonymized student assignments are included on the companion website.
11 All student names have been changed to respect student privacy.

12 On the complexity of kinship, see Max Liboiron's conversation with Rick Harp and Candis Callison on two episodes of the podcast *Media Indigena*, "Pollution is Colonialism: Part 1 (Ep. 258)," May 27, 2021 https://mediaindigena.libsyn.com/pollution-is-colonialism-part-1-ep-258 and "Pollution is Colonialism: Part 2 (Ep. 259)," May 29, 2021, https://mediaindigena.libsyn.com/pollution-is-colonialism-part-two-ep-259, accessed June 17, 2021. See also Liboiron (2021, esp. p.109), Todd (2017), and Wilson (2008).

References

Barker, J. 2018. Territory as analytic: The dispossession of Lenapehoking and the subprime crisis. *Social Text*, 36 (2), 19–39.

Berlant, L., 2016. The commons: Infrastructures for troubling times. *Environment and Planning D: Society & Space*, 34 (3), 393–419.

Bjornerud, M., 2018. *Timefulness: How thinking like a geologist can help save the world*. Princeton, NJ: Princeton University Press.

Conquergood, D., 2002. Performance studies: Interventions and radical research. *TDR/The Drama Review*, 46 (2), 145–156.

Cordova, V. F., 2007. *How it is: the Native American philosophy of V. F. Cordova*. Edited by Kathleen Dean Moore, Kurt Peters, Ted Jojola, and Amber Lacy. Tucson, AZ: University of Arizona Press.

Dei, G. J. S., and Jajj, M., 2018. Introduction. In G.J.S. Dei and M. Jajj, eds. *Knowledge and decolonial politics: A critical reader*. Leiden: Brill, 1–10.

Feld, S., 2015. I hate "sound studies". Retrieved from https://steven-feld-a936.squarespace.com/s/I-Hate.pdf.

Freeman, E., 2010. *Time binds: Queer temporalities, queer histories*. Durham, NC: Duke University Press.

Glissant, É., 1997 [1990]. *Poetics of relation*. Translated by B. Wing. Ann Arbor, MI: University of Michigan Press.

Halberstam, J., 2020. *Wild things: The disorder of desire*. Durham, NC: Duke University Press.

Haraway, D. J., 2016. Staying with the trouble: Anthropocene, Capitalocene, Chthulucene. In J. W. Moore, ed. *Anthropocene or Capitalocene? Nature, history, and the crisis of capitalism*. Oakland, CA: PM Press, 34–76.

Hecht, G., 2018. Interscalar vehicles for an African Anthropocene: On waste, temporality, and violence. *Cultural Anthropology*, 33 (1): 109–141.

Hurley, J., 2020. *Infrastructures of apocalypse: American literature and the nuclear complex*. Minneapolis, MN: University of Minnesota Press.

Jackson, Z. I., 2020. *Becoming human: Matter and meaning in an antiblack world*. New York: New York University Press.

Keeling, K., 2019. *Queer times, Black futures*. New York: New York University Press.

Kelley, R. D. G., 2002. *Freedom dreams: The Black radical imagination*. Boston, MA: Beacon Press.

Kohn, E. 2013. *How forests think: Toward an anthropology beyond the human*. Berkeley, CA: University of California Press.

Liboiron, M., 2021. *Pollution is colonialism*. Durham, NC: Duke University Press.

Latour, B., 1993. *We have never been modern*. Cambridge, MA: Harvard University Press.

Lubrano, A., and Gammage, J., 2019. Study: Philly among leaders in gentrification, which has pushed out people of color. *The Philadelphia Inquirer*, 20 March. Retrieved from www.inquirer.com/news/gentrification-philadelphia-african-american-latino-investment-neighborhood-20190320.html.

Mignolo, W. D., 2011. *The darker side of Western modernity: Global futures, decolonial options*. Durham, NC: Duke University Press.

Moreton-Robinson, A., 2015. *The white possessive: Property, power, and Indigenous sovereignty*. Minneapolis, MN: University of Minnesota Press.

Niess, A., 2021. Breath-controlled instrument design for ecological crises. Master's thesis, University of Pennsylvania. Retrieved from https://repository.upenn.edu/edissertations/3686.

Niess, A. 2022. Reorienting sonic creativity amid ecological disorientation. PhD diss., University of Pennsylvania.

Nixon, R., 2011. *Slow violence and the environmentalism of the poor*. Cambridge, MA: Harvard University Press.

Reed-Danahay, D., 2017. Autoethnography. Retrieved from http://dx.doi.org/10.1093/OBO/9780199766567-0162 (accessed January 2, 2022).

Richardson, J., Mitchell, B., and Franco, J., 2019. Shifting neighborhoods: Gentrification and cultural displacement in American cities. Retrieved from https://ncrc.org/gentrification/.

Rifkin, M. 2017. *Beyond settler time: Temporal sovereignty and Indigenous self-determination*. Durham, NC: Duke University Press.

Smucker, J. M., 2017. *Hegemony how-to: A roadmap for radicals*. Chico, CA: AK Press.

Sze, J., 2020. *Environmental justice in a moment of danger*. Oakland, CA: University of California Press.

Sze, J., 2016. Scale. In J. Adamson*et al.*, eds. *Keywords for environmental studies*. New York: New York University Press, 178–180.

TallBear, K., 2015. An Indigenous reflection on working beyond the human/not human. *GLQ: A Journal of Lesbian and Gay Studies*, 21 (2–3), 230–235.

TallBear, K., 2018a. Making love beyond settler sex and family. In A. Clarke and D. Haraway, eds. *Making kin not population: Reconceiving generations*. Chicago, IL: Prickly Paradigm Press, 145–164.

TallBear, K., 2018b. Yes, your pleasure! Yes, self-love! And don't forget, settler sex is a structure. The Critical Polyamorist (blog). April 22. Retrieved from www.criticalpolyamorist.com/homeblog/archives/04-2018.

TallBear, K., 2019. Caretaking relations, not American dreaming. *Kalfou*, 6 (1), 24–41.

Todd, Z., 2017. Fish, kin and hope: Tending to water violations in *amiskwaciwâskahikan* and Treaty Six territory. *Afterall*, 43, 102–107.

Tomlinson, G., 2016. Sign, affect, and musicking before the human. *Boundary* 2, 43 (1): 143–172.

Vázquez, R., 2017. Modernity coloniality and visibility: The politics of time. *Sociological Research Online*, 14 (4): 1–7.

Wilson, S., 2008. *Research is ceremony: Indigenous research methods*. Black Point, Nova Scotia: Fernwood.

Wynter, S. 2003. Unsettling the coloniality of being/power/truth/freedom: Towards the human, after Man, its overrepresentation—an argument. *CR: The New Centennial Review*, 3 (3). 257–337.

3

FIELD JOURNALING IN THE WILD

Defamiliarizing Everyday Environments in Environmental Humanities Courses

Summer Harrison

The retention challenges faced by colleges like mine, which lose 10–25 percent of students before their sophomore years (US News 2020), have led to an increasing demand for programs to deliver "high impact" experiential learning opportunities even as budgets are stretched ever thinner. At the same time, students expect their coursework to meet needs beyond strictly disciplinary knowledge and skills. A UCLA study found, for example, that about two-thirds of students rated it "essential" or "very important" for their college to help them "enhance their self-understanding (69%)," "develop their personal values (67%)," and "provide for their emotional development (63%)" (Astin et al. 2007). Faced with these pressures, an environmental humanities instructor taking a class outside to contemplate a tree, write about a park, or walk meditatively around campus can seem frivolous at best and a waste of precious learning time at worst. Yet, recent research and my own experience suggest that such field experiences, when properly scaffolded and critically approached, provide opportunities not only for high impact student learning but also for valuable student reflection and growth.[1]

My first semester teaching environmental literature at a small liberal arts college highlighted fieldwork's transformative potential. When I asked my students, many of whom were environmental studies and sciences majors from the area, to reflect on local landscapes, many characterized them as "boring" or "uninspiring." Even those who described suburban New Jersey as "pretty," "homey," or even "beautiful" often saw in it not awe-inspiring ecological treasures but just "plain old" woods, swamps, and hills. Other students imagined embarking on environmental careers "out west" where the "real" wilderness is, considering the NJ/NY metro area too densely populated and domesticated to be "real" in this sense.

When I took them to the campus arboretum, where sits a large sign that misquotes Henry David Thoreau's famous line, "in wildness is the preservation of the world" (see Figure 3.1), I was unsurprised to find that most students did not

DOI: 10.4324/9781003221807-3

consider these familiar 16 acres to be a wild space, much less a wilderness. As we sat on the ground among the trees deliberating the rhetorical and historical differences between "wilderness" and "wildness," something about inhabiting a space marked so plainly by these philosophical tensions resulted in a different quality of investment in the ethical stakes at hand. This discussion turned out to be the beginning of a semester-long inquiry about the value of reframing even familiar or mundane spaces as wild—that is, coming to notice the wildness in them, by which I mean the nonhuman agency, the strangeness, the hidden palimpsests of power and injustice, and the layers of personal and cultural stories that are under the surface of what we ordinarily perceive. As a teacher, I was inspired by this experience to investigate how a pedagogy that incorporates direct experience in the field can effectively defamiliarize everyday environments to spark a deep curiosity and break down problematic assumptions about what counts as "wild."

FIGURE 3.1 Sign at the entrance of Drew University's Zuck Arboretum.
Source: Summer Harrison.

While I have experimented with a range of campus-based fieldwork in my environmental humanities courses, I have found a field journaling assignment to be most beneficial for meeting this goal. Students respond to a sequence of activities and prompts that first encourage them to defamiliarize place on a sensory level, through meditative contemplation and bodily awareness, and then on a structural level, through engaging with environmental and political complexities of familiar spaces. This assignment not only establishes a regular, low-stakes writing practice that develops students' habits of observation and reflection but also helps them to see once-familiar locations in more "wild" and complex ways—prompting new insight into their relationship with the world and unraveling persistent assumptions about their surroundings. As students realize how becoming more attentive to the richness of overlooked environments affords a unique pleasure, much like close reading a text, they often develop personal connections to environments that drive curiosity and an internal motivation to learn. The journaling work also encourages students to critically interrogate the ways cultural, literary, and political narratives shape our interpretations of the world, often unconsciously, as well as to consider how we might change these relationships.

Perhaps the most obvious benefit of campus-based fieldwork is its ready access, but there are also other ethical and practical reasons that campus environments should not be overlooked. This work allows students to encounter wildness in what Bill Cronon (1995) calls the "middle ground" of where we actually live. Ignoring these familiar places in favor of more exotic wilderness reinforces an environmentally destructive dualism separating us from "wild" nature and rendering often violent histories of conflict invisible. Echoing my own experiences in the classroom, field pedagogy researchers Goralnik et al. (2012, p. 422) found that "closer inspection of students' home terrain can enliven their curiosity about the wonder and complexity that surrounds them in their everyday environments, rather than reinforce the notion that environmental learning and ethics apply only to special faraway places." Furthermore, in a higher education landscape of increasingly strained budgets, fieldwork based on campus or nearby can meet the need for more high-impact experiential learning at low or no cost, allowing its use in an expanded range of courses. Field courses that integrate campus-based content can also be useful for "further[ing] [students'] sense of belonging" at a school, potentially aiding in both retention and flourishing (Galle 2017, p. 98).

Field Journaling Assignment

In the sections that follow, I describe the assignment's design, focusing on a series of journal-producing fieldwork sessions that draw on insights from contemplative pedagogies and place-based education (PBE), and then conclude with suggestions for successfully incorporating such fieldwork into an environmental humanities course. My semester-long field journaling assignment was developed in "Nature Writing," a 200-level hybrid literature-creative writing course in which students encounter not only classic examples like Thoreau but also explore nature writing's relationship to power and (in)justice through a range of nineteenth- to twenty-first-century works by mostly US authors. The assignment asks students to participate in fieldwork sessions and reflective exercises designed to inspire short (400–700 word) journal entries. To impart a ritual regularity, each entry begins by recording basic information (date, time, place, weather) and incorporates three fundamental elements: observation (of, say, a particular place on campus or a park or building nearby), reflection (on how the observation relates to a larger concept, theme, personal or cultural insight, or reading from class), and a visual (usually an in-situ sketch or smartphone photo). While most students use paper notebooks, some employ digital journaling apps, like Penzu or DayOne, or even a simple file of Word documents. Students write at least ten of these journals throughout the fourteen weeks of the semester and work up a collection of the best four to submit for a holistic grade at the end of term. I strongly encourage students to make journaling a regular part of their day or week, and most, once they get in the habit of it, end up writing many more than the minimum requirement of entries.

While the content of journals may be inspired by a variety of sources—readings, discussions, or random moments of observation—as well as the assignment's own prompts and activities, I will focus here on the series of in-class fieldwork sessions I use to scaffold the assignment's underlying goal of defamiliarizing everyday environments. I divide these into three emphases that build on one another: contemplative practices, critical fieldwork, and expanding place.

Contemplative Practices

We begin the semester by working on close observation, a capacity crucial to successful field journaling. Students learn to defamiliarize their everyday environments through activities that help them recognize how interiority affects our experience of the exterior world. These sessions draw on emerging research in contemplative pedagogies, which use practices such as guided sensory exercises, meditative walking, and moments of silence to nurture "awareness, concentration, insight, and compassion" (Owen-Smith 2018, p. 24).[2]

Despite a recent "contemplative turn," these practices are still underused in higher education, especially in the humanities classroom, and even in PBE programs. Research suggests that teaching the "whole person" can "enhance student attention, cognition, emotional wellbeing, and creativity," and "foster new skills for self-awareness, tolerating intellectual and emotional ambiguity, embracing diversity, civic discourse, and collaborative action" (Litfin 2020, p. 57). Furthermore, because it can stimulate for students "profound inquiries into the nature of themselves and the world around them" (Barbezat and Bush 2013, p. 20), contemplative education may be especially suited to environmental humanities courses which often seek those same outcomes. Writing courses in particular, research suggests, benefit from "contemplative writing" practices, such as the reflective field journaling in this assignment, because they "remind us of the significance of pondering, noticing, and stillness for effective writing … [To] write well, students need the time to explore both their inner and their external worlds" (Owen-Smith 2018, p. 51). Contemplative practices may also be specifically critical for place-based pedagogies, which, despite their theoretical acknowledgement of the importance of students' inner lives to learning, have often suffered from an "absence of interiority" remedied by a "contemplative approach to PBE" (Owen-Smith 2017, p. 24). Exploring interiority alongside the external natural world defamiliarizes students' surface perceptions of their environments and helps them gain a richer understanding of how their own feelings, assumptions, values, and bodies structure these perceptions.

Below are two examples of contemplative fieldwork sessions that directly inspire journal entries while also building fundamental skills of observation and attention. Before embarking on these sessions, students learn the basics of mindfulness, practicing present moment attention (most often to breath) and nonjudgmental return to focus during brief pauses integrated into most class periods.

- *Sensory awareness*: After reading Mary Oliver's "The Summer Day" or John Burroughs's "The Art of Seeing Things," we do a "separate senses" session. Students go outside and cover their eyes to focus on sound, cover their ears to focus on sight, and cover both to focus on touch and smell—prompting acute present-moment awareness of sensory details normally overlooked in common spaces.[3] Building on Oliver's and Burroughs's writing, students then journal on what they had not noticed before and reflect on why this kind of noticing might be, in Oliver's words, a worthwhile thing to spend one's "wild and precious" time doing.
- *Meditative movement*: After reading Thoreau's and Rebecca Solnit's theories of walking, we undertake a walking meditation either at the campus labyrinth or in any nearby location with sufficient space. Students can opt to do a "walking meditation" or a "thought ramble" journal entry. For the walking meditation, they focus either on their own bodily sensations (the feeling of pressure on the foot) or on the present physical world around them (sensory details). For the thought ramble, students walk slowly and randomly while contemplating a chosen question (for Thoreau and Solnit it was the value of walking itself) and refocusing their mind as it wanders. The activity and its subsequent journaling attune students to the present, somatic experience of place and to ways they might use physical movement in their creative process.[4]

Critical Fieldwork

Building on this contemplative foundation, the assignment's subsequent "critical" fieldwork sessions encourage students to further defamiliarize everyday environments by engaging with their environmental, cultural, and political complexities. Uncritical models of PBE have been historically problematic because of their failure to politicize place—that is, their tendencies to idealize certain places, to privilege the local over the global and the rural over the urban, and to ignore the impact of violent colonial histories and relations of race, class, gender, and ability on place. This assignment draws on insights from various critical approaches to fieldwork. Gruenewald's (2003, p. 9) "critical pedagogy of place," for instance, proposes dual goals of "reinhabitation" or learning to "live well" in "places that have been disrupted or injured," and "decolonization," or learning to "recognize" and "address" the causes of this harm. The defamiliarization aim in my assignment is akin to this "decolonization" in that it emphasizes recognizing "ways of thinking that injure and exploit other people and places," an oft-neglected focus in more historically "reinhabitation-centered" PBE models. Likewise, by having students consider place as continually constructed through competing narratives that link the local and extralocal, I draw on Renshaw and Tooth's (2019, p. 2) "place-responsive pedagogy," the goal of which is "understand[ing] the cultural and material complexity of place and the ethical responsibility we share to care for local places in order to address global challenges."[5] By framing our fieldwork critically, we can avoid erasing histories of conflict or perpetuating romanticized fantasies of nature over concerns like environmental justice.

Though most prominent in science, service learning, and travel courses, place-based fieldwork is particularly well-suited to humanities settings because of our skill in dealing both with political tensions and with representation more generally. If, as Somerville (2010, p. 342) writes, "our relationship to place is constituted in stories," a goal of critical place-based pedagogy in the humanities might be to uncover hidden stories of place to generate new narrative relationships. In engaging multiple literary texts alongside fieldwork, we defamiliarize assumptions of place, or "nature" itself, as a single stable narrative, revealing places as palimpsestic in their meanings, histories, and relationships to power. This kind of critical fieldwork attentive to storied notions of place can "enrich student understanding of course materials," thereby "deepening the ethical dimensions of environmental inquiry" (Atkinson 2013, p. 255), and "move students to reflect on their own cultural assumptions and shift their perspectives and values" (Renshaw and Tooth 2019, p. 14). In my own experience, students have been much more successful in gaining a complex, storied understanding of place, and of their own relation to places, when we pair texts and fieldwork rather than approach either alone. Below are examples of sessions that have successfully inspired critical journal entries.

- *Personal and cultural place*: To prompt awareness of how personal and cultural histories of power and otherness shape students' encounters with the natural world, students revisit a significant childhood place (in person or imagination) to reflect on how their own background has shaped their notion of the wild. Evelyn White's "Black Women and the Wilderness," about the lifelong effect of childhood racial violence on Black women's experience of nature, is an ideal pairing text (White 1991). Then I expand this personal focus to the campus and surrounding town. Students act like investigators, uncovering town historical markers and engaging with its past including the Lenape people's use of the land and displacement, and how the railroad near campus affected the urban/suburban dynamic and local rose industry. These activities lead students to consider questions like, "how did we come to believe that places could be 'owned' and whose bodies and knowledges have been brutalized in the process, rendered illegible, and/or invisible?" (Shannon and Galle 2017, p. 4).[6] They also emphasize for students the dynamic construction of place through competing and overlapping narratives, both suggesting that places are not timeless, stable entities and highlighting, as they write about them, the ethical ramifications for their own participation in this long history of place representation.

- *Food*: To consider access to nature as food, students visit and compare locations like the campus dining hall and community garden, a nearby urban farm and a bodega, heightening their awareness of the ethical, aesthetic, and environmental stakes of food production/consumption. I often pair this with Helena Viramontes's novel *Under the Feet of Jesus*, about a Mexican-American family of migrant grape pickers, and a mindful raisin eating meditation (Viramontes 1995).[7] Such activities defamiliarize the mundane act of eating, attuning students to the often unnoticed sensory and political dimensions of their everyday food.
- *Natural resources*: Energy's invisibility and implications for both the environment and human health take on new meaning after we walk around observing the way our campus uses energy before discussing Ann Pancake's *Strange as This Weather Has Been*, a novel about mountaintop removal coal mining in Appalachia (Pancake 2007). As part of this session, I pinpoint exactly which coal operations are currently supplying campus power, rendering the invisible interconnectedness between places more concrete. Alongside Linda Hogan's *Solar Storms*, an environmental justice novel about the effect of hydropower development on Indigenous populations (Hogan 1997), we take an out-of-class walking fieldtrip to the nearby Loantaka Brook (tributary of the Passaic River) to consider the role of hydropower in our area, including the impact on the Lenape peoples.
- *Nonhuman personhood*: Unsettling ingrained anthropocentric assumptions can be powerfully defamiliarizing. To do this we read Robin Wall Kimmerer's (2017) "Learning the Grammar of Animacy" and follow her advice to practice interacting with nonhuman entities as persons—for instance, engaging trees as sentient "standing people" to be conversed with rather than either objectified or possessed—as a way to "remind us of the capacity of others as our teachers, as holders of knowledge, as guides" (Kimmerer 2017, p. 58). Complementary sessions can be spent on Traditional Ecological Knowledge (TEK), the legal status of nonhuman entities (e.g., Maori granting of rights to Whanganui River), and nonwestern models of personhood/agency.
- *Suburban and urban nature*: After reading Jennifer Price's "The Plastic Pink Flamingo: A Natural History" (a chapter in Price 2017), students bring in an array of found lawn ornaments to create a temporary art installation that interrogates the historical construction of class-based taste in the domestic

landscape. We also defamiliarize the genre by comparing classic nature writing with that based in more urban settings. We might, for instance, sit by a display of campus blooms while comparing the dancing pastoral daffodils of William Wordsworth's "I wandered lonely as a cloud" with G. E. Patterson's brutal critique of unequal urban nature access in "The Natural World."

Expanding Place

Having worked all semester to defamiliarize places, end of term activities ask students to revisit their assumptions of what counts as place in the first place. The value of this approach is underscored in Shannon and Galle's (2017, p.2) recent collection that aims to "engage with, stretch, and play with" the notions of place used in PBE.

- *Body as place*: To defamiliarize our perception of place at different scales, we read selections from Ed Yong's *I Contain Multitudes: The Microbes Within Us and a Grander View of Life* (Yong 2016) and then consider the body itself as a sort of outdoor/indoor landscape complete with different ecosystems. Yong likens inflammatory bowel disease, for instance, to a dying coral reef in which the balance of organisms has altered. This conversation also engages with ways microbes shape our living spaces (e.g., "seeding" buildings with beneficial microbes), and how a vivid awareness of our microbiomes might productively challenge fictions of individuality.
- *Anthropocene place*: Students contemplate the visibility or invisibility of the Anthropocene on campus by creating a collaborative "Cabinet of Curiosities" installation.[8] From dining hall chicken bones (a potential stratigraphic marker of the epoch) to a piece of a campus tree felled by Super Storm Sandy, the curated objects in the exhibit highlight local and planetary connections. With Hurricane Ida we added a muddy doorknob found in the aftermath of flash flooding. Passing students are invited both to suggest additions and to collectively engage with feelings of climate grief and anxiety often ignored in their courses. Such fieldwork may be especially vital for issues like climate change, the largeness and "slowness" of which can make emotional connection and motivation to act challenging.
- *Digital place*: Another way we expand the notion of place is by deconstructing "real"/virtual boundaries. Toward the term's end I pair a reflective (re)visit to the Thoreau arboretum sign with a tour of Walden—not in person but in the open-world simulation *Walden, A Game*, in which players attempt to survive while taking in the "beauty of the woods and Pond, which hold a promise of a sublime life beyond these basic needs" (Fullerton 2017). Thoreau's own penchant for journaling is reflected in the game's journal component, allowing students to write last journals on the act of journaling itself. This game and others, like the online version of the classic Oregon Trail, offer fascinating opportunities to discuss how the construction of physical versus digital places may reify or challenge assumptions about wilderness, race, class, gender, and ability.[9]

Strategies for Success

To conclude, I'll suggest some strategies for successfully introducing fieldwork assignments into environmental humanities courses.

Preparation

Since fieldwork is not typical in humanities courses, I've found it important to emphasize both the work's broader rationale (purpose, goals, benefits) and practical uncertainties (how to prepare for being outside in the weather, safety precautions for solo journal-producing excursions). To clarify the value of contemplative and campus-based fieldwork, I share research on its advantages and encourage students to raise their concerns. Students may doubt "the 'academic' merit of these 'recreational excursions,'" or believe that reflective fieldwork is inherently "sentimental" or diverts energy from the "real work" of "environmental advocacy or scientific inquiry" (Atkinson 2013, p. 257). Yet, as I tell them, research indicates that rather than "compromising the intellectual challenge" of a course, taking class time for such activities can actually help students better "assimilate course material and develop a stronger sense of themselves in relation to it" (Litfin 2020, p. 60). Indeed, while initially skeptical, the majority of my students ultimately report that the sustained contemplation afforded by our fieldwork results in a deeper understanding of both course materials and themselves, making for a particularly invigorating learning experience.

Furthermore, I make explicit the ethical and practical reasons behind a campus-based approach. While students may assume that fieldwork in more exotic or far-away locations is inherently more valuable, an assignment like this offers a great opportunity to explore the significance of defamiliarizing their own campus environment, not least because it gives them a model for contemplatively and critically engaging with other familiar places. In environmental literature courses I also articulate the surprising value, inherent in a campus-based approach, of reading works outside their original settings—of reading Thoreau not at Walden, Muir not at Yosemite. This place-disjointed reading allows students to ponder how the styles and philosophies of such figures continue to shape assumptions about the natural world in unexpected ways—even now, even here.

Regularity

Another strategy for success in my experience is to make fieldwork a regular component of the class, that is, habitual rather than occasional. In "Nature Writing," we leave the classroom to do fieldwork for part of almost every class. Students often use in-class outdoor writing time to spark their thinking for a full journal entry or more formal writing. Course feedback indicates that this regularity is key to helping students develop a consistent writing practice that both increases the amount and quality of their writing and boosts confidence in their identities as writers. Students who were reluctant to participate in fieldwork at first often reflected at the end of the semester that the regularity of the experience got them into a working rhythm, and they ended up really valuing the consistent space to write and reflect, even continuing the practice after the course ended.

Reflection

Finally, I emphasize the importance of integrating explicit space for both group and individual reflection into a fieldwork assignment, during each class as well as cumulatively at the end of the semester. Litfin's (2016, p. 121) phrase for this, which I borrow, is "gathering the harvest." After a contemplative practice or other activity, I ask students to gather the harvest, usually by freewriting or discussing the insights gained—into themselves, others, the world. Often, we use the last 5–10 minutes of class for a guided meditative reflection on the day's subject. At the end of the semester, students reflect collaboratively through the design of a class literary magazine. Each student chooses their favorite example of writing from the class, often a journal entry or a paper inspired by one. After dividing themselves into thematic clusters, each group writes an introductory blurb reflecting on their section, and then curates a few pages of content and images using the free program Lucidpress. Creating the literary magazine has been a productive way to reflect on the semester's work together and gives students a lovely "product" that they are often eager to share with family and friends.

In closing, I reflect myself on that first trip to the arboretum and initial student reluctance to approach everyday local places as meaningfully wild or worthy of sustained contemplation outside of scientific study. I've come to realize that it's not that students don't care about local places but that they need to reacquaint themselves with the wonder and complexity of ordinary wildness by developing contemplative and critical approaches to place. Building these habits and abilities, in part through assignments like field journaling, can engender a sense of empowerment for students as they become more capable of understanding the places in which they find themselves and, thus, better able to imagine change on personal and collective levels. In sum, I propose that a pedagogy designed to defamiliarize everyday environments through fieldwork is both a worthwhile and accessible approach to environmental humanities teaching and learning. We need to recognize, as Thoreau said, that "our schoolhouse is the universe"—and the more we can encourage students to engage with that universe, contemplatively and critically, the better for our environment and one another.

Acknowledgment

We acknowledge that Drew is located in Lenapehoking, the traditional territory of the Lenape people. We offer recognition and respect to Lenape ancestors, peoples today, and the Lenape future to come across Lenapehoking and the Lenape diaspora. In so doing, we acknowledge Lenape Nation sovereignty and their long-standing presence on this land, which precedes the establishment of Drew University, New Jersey State, and the United States of America. We acknowledge and offer deep gratitude to this Lenape land and water that supports us. We invite all members and guests of the Drew community to join in this acknowledgement, respect, and gratitude.

Notes

1 While most of the fieldwork I describe here took place outdoors, I use "field" to refer to any learning location outside the traditional classroom, thus including not just the wilderness or even the outdoors but also built and urban environments.
2 For a comprehensive list of practices, see www.contemplativemind.org/practices/tree.
3 If students have disabilities that might impact participation in fieldwork, we work together in advance to design adaptations that work for them. We also have class discussions about ways traditional environmental discourses of the wild have tended to center white, male, middle- and upper-class, able bodies as the ideal.
4 Some recent place-based pedagogies turn toward the idea of embodiment, of teaching the whole student rather than a "disembodied mind" (Litfin 2016, p. 116). Renshaw and Tooth's "practical materialist approach," for instance, emphasizes that "landscape acts on us and alters our bodies in palpable ways" (Renshaw and Tooth 2019, p. 12).
5 For more on these conflicts in PBE, see also Shannon and Galle (2017), and Somerville (2010).
6 For more on what they call "land education," which emphasizes Indigenous and decolonizing approaches to place-based education, see for example Tuck et al. (2014).
7 I couple a general version of this meditation, focused on close attention to sensory details, with a second post-reading version that brings in more complex considerations of farmworkers' rights and impacts of pesticides. Litfin incorporates a similar session on the "hidden life" of the strawberry in which students discuss farmworker labor, toxic fungal controls, and the energy costs of food production before observing a berry closely, eating it, and reflecting on the experience (Litfin 2016, 129).
8 We model this display on the one at the University of Wisconsin, now documented in a book by Mitman et al. (2018).
9 While PBE scholarship has often been critical of technology as homogenizing or distracting from "real" place, Davis (2017, 181) argues that digital pedagogy and PBE can "mutually benefit each other" since digital simulations allow for student experiences in otherwise inaccessible places.

References

Atkinson, J., 2013. Multi-sensory experience and environmental encounter: Rethinking the sustainability of humanities education. *Interdisciplinary Environmental Review*, 16 (2–4), 253–266.

Astin, A., *et al.*, 2007. The spiritual life of college students: A national study of college students' search for meaning and purpose. Retrieved from https://spirituality.ucla.edu/findings/.

Barbezat, D., and Bush M., 2013. *Contemplative practices in higher education: Powerful methods to transform teaching and learning*. San Francisco, CA: Jossey-Bass.

Center for Contemplative Mind in Society, 2021. Tree of contemplative practices. Retrieved from www.contemplativemind.org/practices/tree.

Cronon, W., 1995. The trouble with wilderness; Or, getting back to the wrong nature. In W. Cronon ed., *Uncommon ground: Rethinking the human place in nature*. New York: Norton, 69–90.

Davis, R., 2017. Redefining learning places in the emerging digital ecosystem. In D. Shannon and J. Galle, eds., *Interdisciplinary approaches to pedagogy and place-based education: From abstract to the quotidian*, London: Palgrave Macmillan, 167–190.

Fullerton, T., 2017. Walden, a game. University of Southern California Game Innovation Lab.

Galle, J., 2017. The potential for place-based learning experiences on the college campus. In D. Shannon and J. Galle, eds., *Interdisciplinary approaches to pedagogy and place-based education: From abstract to the quotidian*, London: Palgrave Macmillan, 85–104.

Goralnik, L., *et al.*, 2012. An environmental pedagogy of care: Emotion, relationships, and experience in higher education ethics learning. *Journal of Experiential Education*, 35 (3), 412–428.

Gruenewald, D., 2003. The best of both worlds: A critical pedagogy of place. *Educational Researcher*, 32 (4), 3–12.

Hogan, L., 1997. Solar storms. New York: Scribner.

Kimmerer, R. W., 2017. Learning the grammar of animacy. *Anthropology of Consciousness* 28 (2), 128–134.

Litfin, K., 2016. Person/planet politics: Contemplative pedagogies for a new earth. In S. Nicholson and S. Jinnah, eds., *New earth politics: Lessons from the Anthropocene*, 115–134.

Litfin, K., 2020. The contemplative pause: Insights for teaching politics in turbulent times, *Journal of Political Science Education*, 16 (1), 57–66.

Mitman, G., Marco, A., and Emmett, R., eds., 2018. *Future remains: A cabinet of curiosities for the Anthropocene*, Chicago, IL: University of Chicago Press.

Owen-Smith, P., 2018. *The contemplative mind in the scholarship of teaching and learning*, Bloomington: Indiana University Press.

Owen-Smith, P., 2017. Reclaiming interiority as place and practice. In D. Shannon and J. Galle, eds., *Interdisciplinary approaches to pedagogy and place-based education: From abstract to the quotidian*, London: Palgrave Macmillan, 23–36.

Pancake, A., 2007. Strange as this weather has been. Berkeley, CA: Shoemaker and Hoard.

Price, J., 2017. Flight maps: Adventures with nature in modern America. New York: Basic Books.

Renshaw, P., and Tooth, R., 2019. Diverse place-responsive pedagogies: Historical, professional, and theoretical threads. In P. Renshaw and R. Tooth, eds., *Diverse pedagogies of place: Educating students in and for local and global environments*, New York: Routledge, 1–21.

Shannon, D., and Galle, J., 2017. Where we are: Place, pedagogy, and the outer limits. In D. Shannon and J. Galle, eds., *Interdisciplinary approaches to pedagogy and place-based education: From abstract to the quotidian*, London: Palgrave Macmillan, 1–10.

Somerville, M., 2010. A place pedagogy for "global contemporaneity". *Educational Philosophy and Theory*, 42 (3), 326–344.

Tuck, E., McKenzie, M., and McCoy, K., 2014. Land education: Indigenous, post-colonial, and decolonizing perspectives on place and environmental education research. *Environmental Education Research*, 20 (1), 1–23.

US News, 2020. Freshman retention rate: National liberal arts colleges Retrieved from www.usnews.com/best-colleges/rankings/national-liberal-arts-colleges/freshmen-least-most-likely-return.

Viramontes, H., 1995. Under the feet of Jesus. New York: Penguin.

White, E., 1991. Black women and the wilderness. *Outdoor Woman*, January.

Yong, E., 2016. I contain multitudes: The microbes within us and a grander view of life. New York: HarperCollins.

4

GO BOLDLY!

Empowering Students to Find their Stories in the Wild

Ellen Bayer

Wanting the Wild

The Pacific Northwest might call to mind stereotypical images of puffy jacket clad and hiking booted folks tromping through evergreen forests and summiting jagged peaks. To be sure, there's a lot of that here, but there's also a significant portion of the population who does not enjoy such easy access to our multitude of national parks, wilderness areas, and other public lands. What I quickly came to learn is that many of my students fall into the latter group. Many of them have never been on a hike, much less set foot on the snowy flanks of the nearby stratovolcano, Tahoma, whose glaciated crown looms over our campus.

There is a hunger for outdoor adventure, but many of my students must navigate substantial challenges to make that possible. These challenges range from such things as not having a vehicle with which to access trailheads; lack of information, such as not knowing what type of pass or permit is required for certain locations; leading full time lives of work and family care on top of being full time students; feeling like the natural world is not a place where they belong; fear of an unknown environment and of the potential for violence there; limited resources, not just financial but also in terms of gear, clothing, and leisure time; and not having a community with whom to share these experiences. Such obstacles can be overwhelming and intimidating, and they are shared by students in urban and rural communities alike.

In response, I developed several assignments that facilitate and support students' personal journeys into the natural world. One that I outline here is a scaffolded assignment that combines a nature excursion with live storytelling, and it provides students an opportunity to practice and hone their analytical and creative skills while folding that into experiential learning. We begin by listening to and analyzing a range of outdoor adventure stories, which introduces students to the genre and equips them to engage with stories from an analytical perspective. Next,

DOI: 10.4324/9781003221807-4

students embark on an excursion of their choice in the natural world or attend one that I organize. Finally, students develop and then share their own stories of outdoor adventure during a live storytelling event at the end of the term. They leave the class with at least one foray into the woods under their belts and with the assurance that their experiences are story worthy.

Learning The Art of Live Storytelling

The inspiration for the storytelling assignment was *Boldly Went*, an outdoor adventure storytelling podcast. *Boldly Went*'s core values include proactively fostering inclusivity and equity; supporting local economies; and practicing environmental stewardship, which made it a good fit for my introductory environmental literature course. The podcast features a range of storytellers from diverse backgrounds and varied experiences with adventuring in the outdoors. I make a point of including stories that aren't about epic adventures told by hardcore outdoorspeople. This allows students to hear stories from people with whom they could relate or identify. Additionally, I weave in stories told by previous students, giving the current class an opportunity to hear from their peers. Unfortunately, the *Boldly Went* podcast has ended, but some episodes are still available on Soundcloud.[1] There is a plethora of other live storytelling podcasts out there (*The Moth* is a particularly useful example), and instructors might choose to incorporate other titles into their courses to use as models.

We begin most class sessions by listening to one story, either from the podcast or from one of my former students. Before listening, we refresh the set of criteria that we'll use to frame our discussion. These criteria, adapted from *Boldly Went*, outline the key characteristics of a strong story:[2]

- *Go out hard and finish strong*: Does the story have a compelling first and last line to strongly start and end with? What grabbed your attention?
- *Stay on course*: Does the story share a brief overview of the adventure then craft a story that creates suspense and keeps the listener's attention? Does it keep the logistical details brief and place emphasis instead on the story?
- *Identify the stakes*: Does the storyteller have something to gain, lose, or risk in this adventure? The stakes don't necessarily have to be high, but the storyteller should show some type of change, even if small.
- *Evoke emotions*: Does the story make you want to laugh, cry, or hurl? How does the story pivot on emotion?
- *Provide a takeaway*: What is the key learning point or moral of the story? This might be implicit in the details, or the storyteller might directly state it. The point may be minor or major.

After listening to the story, students have five minutes to free write a response, using the criteria outlined above. From there, we work through each criterion and conduct an analysis of the story as a class. I let students know in advance that everyone will have to share one thought; we begin with volunteers and end by going around to anyone who had yet to contribute. This helps to reinforce that they all have insights to share and that they each contribute to our understanding of the story and its craft. Using the criteria as a framework ensures that we maintain an analytical focus instead of veering into an evaluation of the story and its teller. This, in turn, reassures students that when it comes time to tell their own story, the audience won't be judging them.

It doesn't take long for students to become quite adept at analyzing the stories, and they come to see what makes a story work. Furthermore, they can also identify areas for improvement while also noting ways a storyteller might execute such revisions. This practice not only equips them with foundational analytical skills, but it also prepares them to craft their own stories and provide useful feedback when conducting peer review later in the term.

To hold to my belief that I shouldn't ask students to do anything I wouldn't do myself, I tell a story early on during the term. In this way, I'm the first one to make myself vulnerable by standing in front of them and sharing my personal experience. We use the same criteria for analysis, and students have offered me some great insights into my own work.

Venturing Out

With the storytelling groundwork laid, we transition to students embarking on their own nature excursions and developing their stories. The excursion necessitates

FIGURE 4.1 Students hiking the Pacific Crest Trail near Chinook Pass, Washington. Source: Ellen Bayer.

some additional guidance on the part of the instructor. It's important to notify students about this component of the course from the first day so that they can work it into their schedule. Students have the option to plan their own adventure, or they can join one that I organize for the class. For my trip, I list the date on the syllabus so that students have ample time to plan. Students embarking on their own have until the last few weeks of the term to do so.

For my organized trip, I choose locations that students express interest in but haven't had the opportunity to visit on their own. There are some classic hikes in the area that they may have heard of and would like to see for themselves, but we also have some wonderful local parks that can provide an adventure right in our backyard. While it takes more time and resources, I've also worked to secure grants to fund a trip to Mount Rainier National Park, covering transportation, entrance fees, and food. While an expenses-covered trip like this is particularly impactful and more inclusive, I've not yet found a way to provide this on a regular basis. Most terms, I rely on students to carpool as much as possible, or we choose a park in town that public transportation serves. On average, at least a third to half of the class opts to join the excursion I plan. Students have indicated they appreciate going someplace new with me that they might not venture to on their own.

Students who embark on their own adventure have a lot of freedom to define the parameters of their excursion. They can determine what "nature" means to them and how they would like to engage with it. For some, this might mean walking in a local park or even looking for wildness within a built environment, while others might want to go on their first hike, try camping, or stroll along the beach. There's usually a small group of students who want to go solo for the first time, and others prefer to share the experience with family, friends, or classmates. They might have an hour or two between school and work for their excursion, or they might dedicate an entire weekend to it. It's important to allow students the flexibility to design an excursion that fits within their schedule, comfort level, and skill set. Some will choose to challenge themselves, and that will look different for each student. I've seen students who spent an hour in the local park with a parent have just as meaningful an experience as others who have gone bigger and farther. The stories they share become evidence of this.

There is preparatory work to be done to empower students to venture out into what might be a new-to-them environment. On our course website, I provide a host of materials that introduce students to the basics of outdoor engagement, and we talk through it during class. Too much information can be overwhelming, so I choose helpful resources available from sources like the Washington Trails Association and REI that provide user-friendly, easy to digest introductions to venturing outdoors safely.[3] For the trips I organize, students must sign a waiver for the University, but this isn't required if they choose to plan their own excursion. I would encourage instructors to learn the waiver requirements of their own school before implementing an excursion into a course.

Toward the end of the term, students submit an excursion report. This covers the general details of where they went, with whom, why they chose this location,

FIGURE 4.2 Students paddling the Thea Foss Waterway in Tacoma, Washington, with the Port of Tacoma and Tahoma (Mount Rainier) in the background.
Source: Ellen Bayer.

and what they felt about the experience. The report serves as a checkpoint along the way to make sure that students stay on track with the larger assignment and are prepared to draft their story. The report documents the logistical details, which opens space to focus on the story itself as they move forward. It also allows me to check in with students who didn't attend the course trip and to help keep them accountable.[4] While instructors might ask for more substantial proof of an excursion, my approach is to give students the benefit of the doubt.

Drafting the Story

After students return from their excursion, it's time to begin drafting their stories. While I don't tell them this in advance, I give students the option to use a previous experience as the basis for their story. I've found that some students have a story from an earlier excursion that they're excited to tell, and it seems best to give them the creative space to develop a story that really lights their fire. I wait to reveal this option, though, because I want everyone to go into their excursion for the course with the sense that there's a story in every experience.

We begin by grouping into small peer cohorts and talking through their excursions, making note of interesting details and trying to identify a story arc. This is an informal conversation that allows students to receive initial feedback from their peers and to begin laying the foundation for their story. At this point, they might have an idea for the story's takeaway, but this isn't essential. The main objective of this first conversation is to get students brainstorming and to give them an audience to help them begin to tease out the compelling and important details from their experience. Additionally, by keeping the focus on sharing their ideas vocally, instead of in writing, students ease into the realm of live storytelling.

Next, we conduct a brief first and last lines workshop. Students rejoin their small cohorts and share ideas for the first and last lines of their stories. This activity asks students to work out where their story begins and to know where it's going to end. Having a polished first and final line is key to good storytelling, but doing so early in the process encourages students to think through what story they would like to tell and what they intend for the audience to take away from it. This session also provides students an opportunity to check back in with their cohort and talk through any other roadblocks they might be experiencing during the drafting process.

During the early stages of this assignment, I introduce students to a 5 Way Points model of storytelling.[5] This approach employs way points, or chapters, to structure the story. While the way points are unique to each story, there is typically a turn or change that indicates the movement from one way point to the next. The benefit of this model is that it helps the storyteller keep on track; one shouldn't memorize a written version but should instead choose five words or brief phrases that the storyteller then uses to guide themselves through the narrative. Because the storyteller must simply get from Way Point 1 to Way Point 2, and so on, it removes the pressure of memorizing a script verbatim. This, in turn, leads to smoother and more organic storytelling. Furthermore, this organizational scheme forces the storyteller to choose only the essential details—it pares a story down to its core, which will also keep the audience more engaged. As we listen to stories throughout the term, we work on identifying the way points of each one so that students can analyze how a storyteller uses it to unfold their experience. Stories need not follow a chronological order, and using the 5 Way Points model can inspire students to think about the sequence in which they might divulge each new chapter to maintain their audience's attention.

Students conduct a more formal peer review with the rough drafts. While the final draft will be spoken, I ask students to submit a written rough draft. Doing so can help students think through the details of their story in a format that is more familiar to them. It also makes peer review more efficient because students have a tangible document to reference. It's important to instruct students not to get too wedded to the wording of the draft; keeping the emphasis on the story facilitates students learning to tell their story instead of memorizing it. This is essential to strong storytelling and results in a more natural sound to the story. To reinforce this, I design the workshop and corresponding worksheet to encourage students to focus on the core details of their peer's story and not on their phrasing or grammar.[6]

To emphasize further the importance of telling a story instead of reading it, I don't ask students to submit written revisions of their stories. Instead, they take their review feedback into consideration as they polish the final spoken version. Not requiring written revisions encourages students to focus on absorbing their story's essence and to use their way points as a guide instead of trying to recall exact phrases.

Telling the Story

In lieu of a final exam, we hold a live storytelling event. Sharing a story in front of the class is optional, but I offer extra credit for doing so. I have found that, initially, a handful of students will sign up for slots. As I continue to pass around the sign-up sheet, more names are added. On the day of our storytelling event, as students share their stories, some holdouts feel inspired and end up telling theirs live as well. One advantage of the live event is that students will tell their stories instead of reciting or reading them. Again, this leads to more natural storytelling, which is more compelling and engaging. While I have toyed with the idea of requiring everyone to share their story at the event, this voluntary system seems to work quite well and results in most students participating. The event is consistently a course highlight, and the level of engagement, support, and inspiration it generates offers a tangible reward to instructor and students alike.

During the event, I record students' stories using an app, Voice Record Pro. Students simply hold my phone as they speak. Having accidentally missed a couple stories due to not hitting record between storytellers, I now record the entire event as one file. You can use editing software if you'd like to separate the stories later. Voice Record Pro gives you the option to share the files in several platforms, which makes storing them easy and straightforward.

Students who choose not to tell their story during the event need to record it on their own. For some students, it's just too much for them to tell a story in front of the class, so it's important to make this an option.[7] I encourage them to tell their story in front of family or friends so that they have an audience. Many of the students who choose to record their story at home also tend to read their story instead of telling it. While this approach might not be as audibly compelling as the live stories, these students nevertheless develop strong narratives and can still offer thoughtful takeaways.

Navigating Challenges

Instructors will navigate some challenges along the way. In my experience, these tend to fall into three broad categories: time; access; and preparation. I offer here some solutions that have worked in my courses.

Due to the busy lives my students lead, many have claimed not to have time for the excursion. I have learned the value of framing the excursion as an assignment (which, of course, it is). When I introduce the excursion, I emphasize that it is a homework assignment and part of a larger course project. The storytelling project accounts for the largest percentage of the course grade, and students earn a grade for the excursion report (and, by extension, the excursion itself). While the report is a low stakes assignment—they earn credit for submitting it—it nevertheless messages to students that it is something on which they will earn a grade.

Furthermore, I ask students to consider the amount of time they would set aside to develop a term paper. For such a project, they would spend time developing a topic, conducting research, reading their sources, and planning how to integrate that information into their essay—all before even beginning to draft the essay itself. I encourage students to think about their excursion as a comparable information gathering step in the larger project process. We discuss how their experience serves as the foundation for their story, much like primary and secondary sources might be the foundation of a term paper. My sense is students fear that doing something that feels more like fun and less like a rigorous trip to the library isn't a legitimate academic pursuit. To counter this, we discuss the value and purpose of experiential learning and creative scholarship.

Once students feel convinced that they do have time for a foray into the woods, it's important to empower them to access it. This takes several forms. First, as noted above, students may have significant obstacles that can prevent them from accessing parks and public lands. Here in Washington, as in other parts of the United States, many of these places require permits or entry fees, which might be an expense that is beyond the grasp of some students. Many of my students rely on public transportation, which puts most trailheads beyond their reach. Others may not have basic hiking gear, whether that's sturdy hiking shoes, waterproof layers, or something as seemingly simple as gloves.

Before assigning an excursion, it's essential to realistically assess your students' access, or lack thereof, to the outdoors. Remaining open to local parks, accessible by public transportation, as valid locations for excursions is helpful. Offering an instructor-led excursion is another good solution. I can typically cobble together enough students with vehicles so that we can carpool. Some schools have university-owned vans or other vehicles that may be available. National Parks often have entrance-fee waivers for school groups, and you need only submit an application that describes your educational field trip. As a class, we work out gear sharing so that everyone has what they need to venture into the wild. I've found that my outdoor adventure friends have been more than happy to donate used gear for student use as well. On one trip, I brought a bag full of secondhand gloves and hats, which students gladly made use of. I bring some snacks as well so that students who are experiencing food insecurity will stay fueled during our adventure.

Making sure that students are prepared for their excursion is perhaps the most complicated element of this assignment. Many of my students have never been on a hike, so there can be a tremendous learning curve. I make a point of emphasizing that epic adventures are not required for a good story, and I encourage students to stay within their skill set. They might toe the line between comfort and something new or personally challenging, but I remind them that being a little bolder doesn't mean they should leap into something they aren't prepared for. Students seem to take this to heart, and while some certainly have a bigger adventure than they expected, they have consistently made good decisions about what is safely within their wheelhouse. Providing resources with information about adventuring outdoors safely is key, as is taking some time to talk through this in class. An added benefit of the instructor-organized trip is that you have some control over the parameters of the excursion.

Additionally, students may also have a significant fear of the natural world. During the term, I incorporate texts in which writers explore their fears about being in the natural world. This opens important dialogues about the roles that race, gender, trauma, culture, and privilege play in shaping our experiences outdoors. It also opens space for students to reflect on their own fears and invites them to share this with the instructor and their peers. We can then work together toward finding an excursion that will help to mitigate any fears the student may have. Again, the instructor-led trip is a good way to accommodate such situations. Students opting for their own excursions have ventured out with family, friends, and even classmates, which both adds to their comfort and creates a shared experience.

While this project wasn't feasible in the shelter-in-place days of the COVID-19 pandemic, now that we can once again access public lands and parks, it adapts well to the remote learning environment. Many live storytelling podcasts made the shift to Zoom events, and instructors who continue to teach remotely can host their live storytelling event on this or similar platforms.

Noting the Rewards

The rewards of this assignment have outweighed any challenges. Tapping into their creativity, students find their voice through storytelling. The experience invites them to see that their stories, and their voices, matter. Incorporating a creative project into the course provides an enriching way for students to demonstrate what they have learned. I would argue that there's a place for creative projects, and storytelling in particular, across the disciplines; this project would be at home in courses beyond those focused on literature. Students also choose to challenge themselves in some way, whether that is in the excursion or in sharing their story at the live event, and they consistently rise to it.

The assignment provides an opportunity for students to connect with the more-than-human world on their own terms and in a way that can be personally meaningful. Many of my students have reported that the excursion is a transformative experience, and it whets their appetite for future outdoor adventures. Because they come from such diverse backgrounds and lived experiences, students' stories create a broader and more inclusive image of who is an outdoor adventurer, and they present a nuanced collection of stories that illustrate the range of tales we can tell about our experiences in the outdoors.[8] On a larger scale, students can begin to see themselves as stakeholders in addressing the broader environmental crisis. We know that personal engagement with the more-than-human world leads to a sense of personal investment in its conservation and preservation; as humans,

FIGURE 4.3 A student enjoying her first hike, at Mount Rainier National Park. Source: Sushma Chaudhary.

we simply care more if we get to touch the mountain. The aim of this assignment is to equip students with the time and tools necessary to get them there and to create a space for them to share their stories upon their return.

Allow a few of my students to speak for themselves here about the rewards they found in being assigned to venture outdoors:[9]

> Being pushed into a new realm is a decision I rarely make. Fear and doubt always cloud my mind when venturing off into the unknown. To do this [hike] is to show what little bravery I have but exploit my determination and inner strength. A coward in one way but a warrior in another. I don't know a soul that would refer to me as a warrior. Even though that little guy dressed up in a suit of armor is as much a part of me as my heart, bones, and skin, he acted as a gargoyle—showing up to protect only when needed. But for the first time in my life, I went on a hike, a journey into nature itself. I have

dedicated the past three years of my life to studying the environment and sustainability, yet I still had no drive or passion to go out there and see what I had to protect! What I was fighting for! To be given a reason to venture out has given me new life. As cheesy as it sounds, that first hike has made me into a new person. I found enough strength to find that knight and push him out into reality. He became my courage and strength incarnate.

(Tyler)

Exploring nature is something I did not grow up going into or thinking it was meant to explore, it was something I had only seen white people do on television. I had fears of getting kidnapped or murdered while in the process of exploring. I was not afraid of the outdoors, but of the humans there who are able to cause harm, the ones that so easily hurt others, those are the ones that frightened me the most. I knew of the horrendous things humans were capable of and I did not want to be another dead body found. I just needed to find a way to not think about all those unpleasant things. I did not want to think about my family's reaction if something tragic were to happen to me. I had to find a way to stop those thoughts, so I started to jog. That got my heart racing in a good way. I jogged for a while before stopping to catch my breath. This made it easier to forget all about my worries, even for a little while, and it reminded me of the reasons I got into exercise a while back. I started to think about ways to improve myself more, put more time into exercising again and ways to better my overall health. Jogging might not have the best idea since I was a bit out of shape and found it a little hard to breathe, but it made me forget about the scary thoughts, even for a little while.

Being at the trails at that very moment made me not view the world as such a scary place. I saw that trail as a place for me to escape the worries of life, a place where it is in fact safe to be, to let go of all the responsibilities and to just be there. School becomes so burdensome at times, so does work and family life. Having a place to go to alone does not sound like such a bad idea. I did not expect to find comfort and peace at the start of my adventure, but I found so much more. I found the calmness and quietness I was craving for. I had the chance to let go of so many thoughts, people's expectations, my own expectations, and how hard I am on myself. I decided to be gentle and let myself enjoy these small instances of silence. I was scared at the beginning of being alone in an unknown place. Turns out that is just what I needed, I just had to find the real reason of needing that quietness.

(Fatimah)

When this class first started and you asked me if this class would change my mind about being outdoors more, in my mind I laughed and said no. I came home and researched people in wheelchairs who hike and do extreme sports. I found Bob Coomber's YouTube channel and immediately begin to imagine myself on the trail doing what he was doing. What was holding me back?

Myself. My second thought was "I can't," then my brain started to list all the ways I would fail. I watched Bob Coomber's movie, and his opening quote is "get out there and at least try it." Heck yes Bob, SIGN ME UP!

I first took a moment to look at myself in the mirror. I asked myself why I couldn't do what I wanted. I realized that I am a strong independent woman. I can legitimately do all the things I've always dreamed of. I am a Wheelchair-Jedi. I YouTubed wheelchair sports, hiking, kayaking, camping, racing, anything I could think of. It never occurred to me that I could hike the mountains. Like Pete Fromm, I began to get inspired by the people in the books. I now have a bucket list of adventures that I plan to do, kayaking, hiking, hunting, and camping are on that list.

(Leslie)

Sharing My Story of this Assignment, Using the Strong Story Criteria and 5 Way Points

Brittany's Story

"I've looked at that mountain every single day of my life, but I've never been to it. I've never had the chance to touch it." Brittany's voice quavered slightly as she said this, standing up in class and pointing out the window in the direction of Tahoma, the stratovolcano that looms over our campus. Several other students nodded in recognition. I stood dumbfounded, both by this fact and by my own naivety. Within days of moving to Washington, I'd made a bee line to Mount Rainier National Park and marveled at having such a magnificent place right in my backyard. Hearing Brittany's words woke me to the privilege inherent in my Sunday jaunt to the mountain. It's humbling to admit it hadn't occurred to me that local residents who wanted to visit the park might not be able to. I knew in that moment that I wanted to use my course as a vehicle for creating student access to wild places.

Starting with Local Adventures

It started small. In that course, we carpooled to a local park during one class meeting and sauntered along the shores of the Salish Sea. Some students had never been there, despite the park being within the city limits. One brought a friend along to ease her anxiety about participating in this excursion. Before long, she was smiling. The students weren't scaling the mountain, but all of them were delighted to have spent an hour ambling aimlessly, breathing in the sea salt air and watching the tide recede—and getting to count it as schoolwork.

Meeting Wasiq

A few years, and a few local adventures, later, I'd managed to cobble together the means for taking an entire class to Mount Rainier National Park. I applied for an internal grant to cover most expenses, and the park gave us an educational exemption for entrance fees. While I was excited for the class to have this experience, one student stood out as being particularly in need of some mountain time. Wasiq was an Afghan refugee. Though typically silent in class, he came to office hours every day to discuss his work, and his homesickness. He showed me pictures of his home village, nestled in a green valley with the Hindu Kush range rising high above, snow-capped and sublime. Wasiq loved the lush emerald landscape of his adopted home, but he missed the mountains.

Mount Rainier National Park Class Excursion

One of the great moments of my teaching career was seeing the look in Wasiq's eyes as he passed me on the Skyline Trail that loops up Tahoma in Mount Rainier National Park. It was snowing sideways, and the summit wasn't visible through the wall of clouds. Wasiq turned to me, his voice electric, and exclaimed, "I'm going up there! I'm going to the top!" I gently suggested that the few hours we had at the park wouldn't be enough time to reach the summit that was over 8,000 vertical feet above us (not to mention the gear and expertise required). Unphased, Wasiq said he was going for it. I had never seen him look so alive. "OK, just be back at the visitor's center in two hours," I said as he sped ahead.

Wasiq's Story

Wasiq spent considerable time drafting his live story. He never spoke much in class because he was worried about his English, and he didn't plan to share his story at our storytelling event. On the day of the event, a reliably vocal student, Dallas, shared a humorous yet moving story that set the bar high. To my surprise, Wasiq volunteered to go next. He stood and explained his difficult transition to life alone in a new country, so far from family and home. It was stepping foot on the mountains of Washington, he said, that had started to ease his homesickness and make him feel at home here. You could feel students hanging on every word from this silent young man who rarely spoke. When he finished, Wasiq's face beamed with pride, and our admiration rang out in applause. The aim of this assignment is to empower students to find their place, and their voice, in the wild. My heart was full seeing Wasiq had accomplished both.

Acknowledgment

I recognize that all of us at UW Tacoma learn, live, and work on or near the ancestral homeland of the Coast Salish people. In particular, our campus is situated on the traditional territory of the Puyallup. As people on these occupied territories, we have a responsibility to acknowledge the land, the ancestors who have cared for this land since time immemorial, and all our Indigenous connections today. We also have the responsibility to acknowledge the histories of dispossession and forced removal that have allowed for the growth and survival of this nation and institution. In light of this history, let us take active efforts to partner with our Indigenous community members and neighbors to seek justice as we continue our work together as a community of learners, leaders, and educators.

Notes

1 Please see the companion website for sample student stories.
2 I am indebted to Angel and Tim Mathis, co-creators of *Boldly Went*, for their guidance in developing this project.
3 See the companion website for sample introductory materials.
4 Only once—to my knowledge—has a student passed off a previous experience as their excursion for the course. I gently reminded the student that the excursion needed to be taken during the course and gave them a chance to correct this.
5 I am indebted to Dean Burke, master storyteller, for sharing this 5 Way Points model with me and for engaging in many enlightening conversations about the art of storytelling.
6 See the companion website for my peer review guidelines and worksheets.
7 Once, a student didn't tell a story in class and didn't submit his own recording. I learned that he didn't have a cell phone or laptop with which to record it and was embarrassed to say so. We met at the library so that he could use my phone. This experience reminded me to make sure that all students have access to the technology required to complete an assignment. Even if tech is available on campus, students might need help connecting with those resources.
8 Please see the companion website for details on how to contribute your students' stories to this collection.
9 I have changed all students' names in this chapter to protect their privacy.

5

THE WORLD IN A POND

Multispecies Encounters and A Map for Confluent Classrooms

Allison Lee Blyler and Holly Connell Schaaf

In the Amory Woods, a tangle of maple and oak quick with grackles; a pair of mallards paddles in autumnal pools. On the main path a chipmunk, cheeks full, skitters beneath a gazebo that rises from the trees; grass rolls up to the sanctuary's border with the adjacent park. From the main Observation Deck, Hall's Pond is turbid from rain and thick with lilies; the resident great blue heron stalks prey, and a scarlet cardinal flashes through a curtain of willow. On the Upland Path, wild turkeys posture and scratch in leaf litter; in the Formal Garden, rabbits devour an explosion of clover under the Japanese maple. Students watch in pairs at these different points, taking notes on weather, wind, plants, trees, animals, and the traces humans have left: sights, sounds, and smells they will share and incorporate in multiple ways into their coursework.

This chapter explores the sequence of visits to Hall's Pond Sanctuary in Brookline, Massachusetts that form the foundation of our co-taught writing seminar, "The World in a Pond: Observing Nature in Boston and Beyond," and the major assignments directly tied to our students' observations at this urban sanctuary. Specific observational viewpoints at Hall's Pond Sanctuary create a foundational map for our course. These different viewpoints are key to students addressing the challenges of imagining global audiences and seeing from the perspectives of local stakeholders.

We provide here strategies for engaging deeply with a single local site based on the course we co-teach that explores Hall's Pond, which is a 10-minute walk from our Boston University classrooms that line Commonwealth Avenue, the bustling urban street that is our campus. The proximity of Hall's Pond enables our students to visit three times as a class and once on their own, helping them to become patient, focused readers of the fluid and static aspects of its landscapes and to establish a long-term relationship to the place, which makes possible practices of re-reading they can apply to other contexts both inside and beyond the classroom.

DOI: 10.4324/9781003221807-5

Describing coursework and research done at Elon University's protected on-campus forest, Jeffrey Scott Coker (2017, pp. 79–80) discusses "a collaborative pedagogy of place" arguing that "[a]ll of this activity occurring in one place (instead of many) further magnifies the integrative learning and educational impact. Students and faculty are able to make connections across investigations and disciplines that might not be possible otherwise." Although our urban campus means that protecting a 58-acre site like Elon Forest is impossible, sustained exploration of 3.5-acre Hall's Pond enables our students to make multidisciplinary connections of the kind Coker celebrates, revealing that the benefits of focusing on one site close to campus which can be visited during class time do not demand a vast reserve—even a small site inspires a wealth of multidisciplinary discovery.

By focusing on one local place, our course encourages intersectional confluences and connections across cultures and disciplines. We bring together a section of native English-speaking students and a section of international ELL students as part of "Mediated Integration," a curricular initiative of the College of Arts and Sciences Writing Program in which we teach. Mediated Integration was adapted from an original approach, cross-cultural composition, pioneered by Paul Kei Matsuda (Matsuda and Silva 1999), but our approaches can work for any student population. Our pedagogy of place grows from our students' multicultural mosaic of visions, which create a collaborative psychogeographical map of Hall's Pond across observational sessions. Psychogeography, the intersection of psychology and landscape traditionally applied to exploration of marginal aspects of urban spaces, translates here as a collective practice of observation that allows students to see this site as both the sum of its parts and far more than that. Creating a psychogeographical map is essential to cultivating a distinct sense of place that is intensely local and open to larger and more global connections.

Although Hall's Pond's scenery (as Thoreau says of Walden) may seem "on a humble scale," it has a significant history behind it and provides rich opportunities to observe a range of wildlife in an urban matrix, which is increasingly important in local places across the globe due to rapid urbanization. Students who may be well-aware of global threats to wilderness and distant refuges also need a fine-grained local perspective on landscape, species, and diverse human stakeholders so they can see connections and contrasts between local spaces shaped by urbanization but also by the wildlife striving to adapt to these places. Hall's Pond lends itself to these connections: as a sanctuary that almost wasn't, Hall's is the end site of a series of human and nonhuman processes and shifting alliances that have transformed it from White Cedar swamp to a one-acre pond, from public to private to public again. Minna Hall, founder with Harriet Hemenway of the Massachusetts Audubon Society, offered the pond as a gift to the town of Brookline, Massachusetts. It was turned down. The site was slated for development when the Massachusetts Wetlands Protection Act rescued it, and Brookline purchased it as conservation land in 1975, later adding the adjacent Amory Woods and blending these spaces into the sanctuary as we know it today, a dense melding of wood, path, garden, and pond that reflects the processes that formed it and the people, most fervently the Friends of Hall's Pond, who continue to maintain it (Giezentanner and Eunson 1996).

A central goal of our ecopedagogy is to create a confluent classroom that empowers students to discover multispecies and multicultural relationships through sustained engagement with a particular local place—a mission that has taken on more urgency than ever in these post-pandemic times. Focusing on one distinctive local site benefits students by helping them to cultivate "careful and critical attention to the specificity of other species' lifeworlds," a particularly important skill "during an era of escalating change" (Van Dooren et al. 2016, p. 8). In our course, we imagine this critical attention as grounded in a combination of focus and versatility in regard to place—where students move from inside the classroom to the outdoors, visit the same site multiple times with differing viewpoints and observation partners, and wander freely during solo visits. By using data from Hall's Pond for two major projects and responding to reading from multiple disciplines; writing both conceptual and practical arguments; and writing for global audiences who may never have visited the pond and diverse local stakeholders who are deeply invested in the sanctuary, students develop dexterity with regards to both their own relationships to places and to expressing their perspectives on the "lifeworlds" of others.

Pre-reflections and Place Narratives

Our first goal in the class is to establish that we all have relationships to places whether we recognize them consciously or not. Writing and discussing two short assignments about place at the start of the semester enables students to engage with general place-based learning issues. These assignments also help students develop a conscious sense of preconceived ideas that they have about the wilderness, nature, green spaces, and how these perspectives and other past experiences with place will shape their responses to Hall's Pond as an urban sanctuary.

Each student writes and shares with their classmates a place narrative, a prose piece of 400–500 words intended to bring readers to their favorite outdoor place through vivid sensory description.[1] This activity helps students to bond in the classroom over their favorite places and to share compelling experiences in the outdoors. The narratives profile diverse local places from historical monuments and national parks to local hiking trails, backyards, soccer fields, and parking lots. Their chosen sites give students a sense of the range of potential connections to place but also create the opportunity to use examples to discuss patterns in relationships to place that establish a starting point for preparatory discussions before we visit the sanctuary. These discussions encourage students to focus on differences and similarities in their chosen sites and to identify particular tropes and strategies in their descriptions of these unique places such as recounting one strong memory or multiple memories, sharing specific types of sensory experiences, and depicting solitary experiences or interactions with other humans or nonhuman animals.[2] Sharing these narratives also helps us establish a more fundamental facet of the

FIGURE 5.1 Students on the boardwalk in the Amory Woods section of Hall's Pond Sanctuary.
Source: Allison Blyler.

course–-that human beings have deep relationships with places. Reminding students of the potential for these connections helps prime them for developing a significant relationship to Hall's Pond.

Pre-reflections, in which students answer specific questions about how they imagine the sanctuary, are an even more important foundation for observations at Hall's Pond:

1. How do you expect the landscape of Hall's Pond will look? Describe what you think a walk through the sanctuary will be like.
2. Hall's Pond is an urban sanctuary. How do you expect that will affect your experience of nature there?
3. What kinds of animals and plants do you think you might encounter?
4. How do you expect people who visit the sanctuary will behave?

Pre-reflections capture each student's preconceived notions. For example, they often discover that they expected Hall's Pond to be either more urban or less urban, based on the places with which they are most familiar, or they realize that they have not usually considered urban spaces to be natural at all, an insight that later helps spark discussion of course readings and the perspectives of local stakeholders. The pre-reflections are shared upon completion in a Google doc and become the basis for early class discussions about our specific site and required metacognitive assignments later in the semester. Looking back on their pre-reflections after we have visited Hall's Pond, students realize how many more details they notice since they have developed deeper familiarity with the place.

Observations and Notes: Cumulative Effects

At the start of the first visit to Hall's Pond, we give students a tour of the sanctuary so that they have a sense of the whole. We then assign them in pairs to specific observation posts/territories. Having students work with partners helps them observe more details and prepares them to consider audience in their major assignments. They stay at these posts and take notes until we return twenty minutes later to collect them. We scaffold their notes with a handout about observational techniques that directs students to focus on recording what they see, smell, and hear and not to spend too much time interpreting these descriptions while at the pond. We do not tell students how or how much to interact with one another in pairs—they choose whether and when to chat with their partners.[3]

Staying in a specific space enables them to become familiar with the details of the landscape, and it often gives them the chance to focus more closely on the wildlife and observe more diverse behaviors as animals approach them. Placing students in particular spots around the sanctuary also literalizes for them different viewpoints in a helpful way, particularly once we meet back in the classroom after the first visit and subsequent visits.

All students are required to take notes with paper and pencil/pen rather than phones or other devices. Many students could likely take notes more quickly on their phones and some wish they could use their phones to snap a photo of the hunting heron or a swimming turtle, but we encourage the students to use analog methods at the site during the twenty-minute notetaking periods. They are permitted to take photos and videos when we arrive at the sanctuary and at the end of class, but using paper helps focus their attention and hone their observation skills. For ease of notetaking, we allow multilingual students to take their original notes not entirely in English if they so choose. Most students take their handwritten notes in English anyway, but it is helpful for English language learners to be able to quickly jot down in their first languages names of specific plant-life or fast-paced sequences of animal behaviors if they feel they may forget details due to the challenges of translation.

After visits to the sanctuary, students have a day to type up and translate their notes and paste them into the Google doc that we create for each visit. All students then skim the full notes before our next classroom session. Reading their classmates' notes enables them to envision the psychogeographical, temporal map that we as a class are creating. It sparks collective enthusiasm and helps students learn notetaking techniques from each other. Although we provide students with a handout about notetaking and discuss strategies in class before the first visit, seeing a range of examples in their classmates' writing sharpens students' observation and

FIGURE 5.2 Students taking observation notes at Hall's Pond.
Source: Allison Blyler.

organization more precisely as they discover in others' work new motivations for observing, methods for describing and naming landscape features, wildlife, or other phenomena, and new approaches for organizing what they write. As students add to the notes with each visit, they eagerly watch our class data set become deeper and richer over time.

Back in the classroom for the next session, students use these shared notes to compare data in activities that strengthen various skills. For example, they distinguish between details in observation notes that lead to factual questions ("What kind of bird is that?"), that lead to broader philosophical questions ("Does the heron know that humans admire him?"), and that lead to possible arguments ("What does the presence of a warbler suggest about the way in which the sanctuary works as an urban space that supports migratory birds and biodiversity?").[4] Through class exercises, students see a range of contrasts and connections in the style and content of their notes as they recall the different angles from which they were observing Hall's Pond. All activities are designed to ensure that the students engage with the different details in the notes during small group sharing as we circulate around the room and engage with individual groups. We create activities that help students move together through a series of steps that require them to pick out specific details from the whole data set and further refine those selections as they explore complex questions about the site and course readings.[5] These questions spark conceptual and practical arguments of the kind students will craft in major assignments.

During these post-visit sessions, students work in groups of four that consist of two sets of observation pairs. We assign the groups so that some grouped pairs were at adjacent observation posts, others had positions across the pond from each other that could access a partly overlapping view from different angles, and still others did not have a view with any overlap. Through these contrasts, intriguing comparisons develop when we share as a class and map the different pairs' and groups' experiences. All students enjoy different details about the same geese that adjacent pairs experienced and grasp the range of behaviors that can be observed in an encounter with any species. The whole class explores how a Great Blue heron and Double-crested cormorant were affecting each other's behavior based on pairs from across the pond who could each see both birds but who saw one bird more closely than the other. Pairs who had no shared sightlines help everyone realize that even in this small intertwined ecosystem, one pair can get an impressively close view of a red-tailed hawk devouring prey while another can have no idea the hawk was at Hall's Pond at all. These post-visit discussions reveal the value in interdependent perspectives, collective mapping, and shared multispecies encounters.

The sense of varied viewpoints students gain from these discussions generates enthusiasm for changing vantage points during future visits, and as students experience different parts of the sanctuary and team up with different partners, their sense of Hall's Pond grows more vividly nuanced. After two class visits, students wander the sanctuary as they wish during an assigned solo visit. We invite students to share concerns about the solo visit as they would for any disability or access issue so we can provide alternatives. Most students feel safe visiting on their own, but we make sure that the few who do not either visit with a classmate or friend.

These solo visits play multiple roles: they reinforce the students' growing confidence in their own relationships to the site, allow students to apply observational techniques they have learned in class while also giving them the freedom to make choices based on previous encounters, and encourage them to spend time in the landscape as individuals. During this solo visit students are allowed to use phones and take pictures and videos at any time but are still required to use notebooks to record observations. The compilation of solo visits we create and discuss afterwards produces a new kind of map of Hall's: perspectives unfold over a longer period of time—different days, agendas, weather, wildlife.

Multiple visits enable students to perceive changes in the site such as trees turning to autumn color and also features that were always there, but that they did not see at first. The increasing familiarity developed with the site also allows students to notice half-concealed animal residents who they would not have been aware of when first arriving at Hall's Pond. The effects of the set of visits are cumulative, emotional as well as intellectual, and produce cross-cultural, interpersonal, and cross-species affinity and empathy.

As the students move between our inside classroom and Hall's Pond, they also accumulate strategies for analysis, which help to prepare them for their major projects. By exploring their observations together, they are learning from each other with our feedback how to analyze data from their pond expeditions and how to use these data to respond to diverse readings. For example, they practice using their notes as they would other textual evidence, putting their own observations and those of their classmates into conversation with the philosophical perspective of a nature writer, the claims of an ecologist about the value of urban ponds, or an argument against dualism from the fields of environmental history or migration studies.

The observations students conduct at Hall's Pond are the heart of the course, and the work they do with them represents a confluence of processes that happen outside and inside the classroom. The practice they conduct in using their observations is designed to flow seamlessly into the major assignments for the course.

Major Assignments: Global/Conceptual and Local/Practical

When outside-the-classroom experiences are essential to major assignments in the course, students engage more deeply and consistently and in more nuanced ways. In contrast to just having a verbal discussion or requiring students to write about their experiences only in short assignments peripheral to the class, connecting experiential learning to work at the heart of the course encourages students to watch and listen more carefully. In the context of busy student schedules, stress, or distraction, tying observation to major assignments provides initial motivation, but the real goal is to lay a foundation for intrinsic motivation. Even if some students at first sharpen their senses in hopes of strong grades, opening up enables them to establish genuine relationships to each other and the sanctuary, exemplified by a wide array of experiences across time, place, culture, and species. For example, a clutch of students from three different parts of the world exclaimed in surprise over exploding jewelweed pods next to the sanctuary boardwalk, cupping the seeds in their hands, carrying their excitement with them on the walk back to campus that day, and looking for more late pods to pop on our next visit. While these relationships are difficult to quantify, they are, in our experience, the most profound rewards for the students and for us.

The first two class visits and solo visit prepare students for the major conceptual paper for a global audience. For this assignment, students imagine an audience who has never visited Hall's Pond and may be unfamiliar with the specific readings they are using as source, but who has a general interest in the issues they are discussing. Students are permitted to design their own areas of focus involving course readings, including Christopher Hassall's (2014) "The ecology and biodiversity of urban ponds" and John Hultgren's (2015) "Nature, place and the politics of migration," but we give them a range of potential topics to explore, which show them the types of responses to readings and questions the paper involves. Here are a few examples:

1. Hassall discusses habitat connectivity and its effects on biodiversity in urban ponds. Use our observations of Hall's Pond to make an argument that assesses the habitat connectivity of the different parts of the sanctuary and the other

landscapes that border Hall's Pond and analyzes how those different landscapes would be likely to affect the movement of a particular species.

2. Hassall suggests that Nature Reserves are "managed for biodiversity (often birds)" and that "access to public is encouraged but controlled." Through use of our observations and sources about Hall's for background, establish that these two factors are true of Hall's and argue how they affect the interactions of non-human animals and human beings that spend time there.
3. How does rootedness as Hultgren defines it both prevent empathy for others (human or non-humans) and possibly encourage or promote it? Choose several encounters with non-human animals at Hall's Pond and consider what elements of each encounter enable an observer to empathize and what aspects might make empathy difficult or impossible. Does empathy depend on or is it affected by rootedness, a migratory sense of place, or other specific factors between or beyond these two perspectives?

The focuses for this paper give students the chance to analyze data gathered from our local site by engaging with multidisciplinary theories to make global ecological connections.

The major conceptual paper helps students become accustomed to analyzing observations as evidence, a practice with which most have little experience due to the writing curricula at most high schools focusing on text-based literary analysis. Since students generally are more used to doing conceptual papers in high school and in other college writing courses, making this assignment a conceptual paper gives them the chance to get used to the experiential evidence in a format a bit more similar to what they have done in the past. Their direct interactions in real time with human partners on class visits and the note activities help students to be aware of what they must explain to other audiences who have not seen directly what they have observed—their classmates give them a built-in audience. These experiential, person-and-place-driven resources guide their arguments in surprising ways that purely text-driven evidence cannot. Students are at first astonished that their observations can serve as support for arguments and respond to course authors, but as they draft this paper, sharing discoveries from their lived experiences as evidence makes them more invested in their claims because they are active stakeholders. Using their own observations helps to move their analyses away from rote close reading and towards discussion that is more concrete, more independently conceived, and more actively in conversation with multidisciplinary arguments.

The students' experiences at Hall's Pond encourage them to envision audience differently for the Practical Proposal, the final major assignment.[6] In order to present an argument to the Friends of Hall's Pond and other local stakeholders relevant to their chosen focus, students use their observations to propose a beneficial project or set of positive changes to the sanctuary. The contrast between this project and the conceptual/global paper directs students' attention to genre features, as they are required to organize their proposals with sections entitled "Background," "Site Analysis," "Goals," "Implementation," and "Support," a structure that follows many proposals of this kind and is also specifically modeled on structural features of the October 1996 proposal "A Plan for Hall's Pond Sanctuary," available for download on the Friends of Hall's Pond website. We discuss this assignment with students before our third class visit to the sanctuary so they can use this visit to make strategic observations with the assignment fresh in their minds, but the project is a culmination of all their visits.

One of the challenges of this assignment, of course, is narrowing the focus. While this is a common challenge in undergraduate papers, the context of the course and assignment allows us to use place-based methods to help students choose specific focuses based on their observations of human and nonhuman interactions with particular parts or aspects of the sanctuary. They can propose plantings for improving the sanctuary for a specific species or two interconnected species; they can propose new receptacle placements to make it easier for visitors to dispose of trash and recyclables, but they cannot propose digging up the entire sanctuary or diverting all traffic from Beacon Street, the busy thoroughfare that borders one side of the sanctuary, in order to reduce noise. Despite students' initial lack of experience with these types of projects, narrowing practical arguments often becomes easier than refining more abstract arguments because they have so much data at hand about the site and access to knowledge about local stakeholders.

Each student needs to develop practical plans and present them to all relevant stakeholders clearly and enticingly. Students determine and reconcile what different stakeholders would desire—and these desires can be contradictory. Some examples of focuses for student proposals include the following:

- Improving the sanctuary as a habitat for native birds by planting new vegetation in the Formal Garden and tweaking the overall garden design.
- Making Hall's Pond more navigable and more welcoming for people with disabilities by helping Hall's Pond surpass basic adherence to ADA regulations.
- Creating an interdisciplinary two-semester curriculum for students in Boston University's School of Education in which they would learn how to teach about nature to two different grade levels and then partnering with schools close to the sanctuary to put the theory into practice both inside the classroom and outside at Hall's.
- Minimizing the harmful effects of light pollution on the nocturnal ecology of Hall's Pond, particularly mating amphibians, by designing and installing fully shielding streetlights that point downwards to reduce nighttime glare.

These focuses reveal the specificity of successful approaches to this assignment yet also the fact that there are a wealth of different angles students can pursue by focusing primarily on wildlife, on humans, or on particular relationships between the two.

Although it may seem counterintuitive, assigning locally driven projects last gives students the chance to address areas that interest them more thoroughly and precisely by learning how to strategically narrow the angle of their focus (literal observational experiences provide good metaphors for this too). Their practical proposals can have meaningful depth because they have accumulated a semester's worth of data and developed a longer relationship with Hall's Pond. Being able to pitch practical changes requires that students have a meticulous sense of the landscape and also the diverse ways in which human and nonhuman individuals engage with the sanctuary. One of our goals for the assignment is to empower students by helping them realize that they can and should have a stake in what happens in the places they are living. The assignment encourages them to recognize the call to stewardship even if they, like the migratory birds they encounter at the pond and sometimes compare themselves to in discussion, have not lived there their entire lives, as long as they take time to develop intimacy with the places and listen to other local stakeholders standing at different positions around the pond.

Conclusion

A rich and complex local place linked to an intricate network of stakeholders, Hall's Pond Sanctuary provides an ideal site on which to base our confluent classroom. As Shannon and Galle (2017, p. 5) write in their characterization of the origins of place-based education, our pedagogy is "a response to larger concerns about the alienating effects of non-permeable classroom walls that [separate] students from their communities rather than embedding them within the ecologies where they are located." Hall's historical fluidity, biodiversity, and complex ecosystem and landscape inspire our pedagogy in this particular course, but one significant takeaway for us is that local-global connections and a direct flow from outdoors to traditional learning spaces can be created from many different kinds of sites. It is this kind of confluence that we hope to create and present as a template for others.

The ongoing collective mapping based on observation that we find so valuable is universally applicable and helps to create a sense of stewardship that goes beyond the classroom. Whether students are observing city streets and parking lots, farmland, backyards, public gardens, pocket parks, neighborhoods, or wildlife sanctuaries, visiting a single site multiple times can establish distinct viewpoints and a collective perspective—and encourage purposeful re-reading that can transfer to other domains. Every site, even if it is small, can generate many possibilities for discovery across disciplines, opportunities to see how local places are shaped by global environmental challenges, and provide an array of lifeworlds for students to observe and learn from.

Even on campuses where a suitable site close enough to bring students during class periods does not seem apparent, local research can reveal unexpected gems—we taught for years near Hall's before discovering it. But multiple visits to a local place too distant to reach during class can provide similar benefits as students can still engage in pre-reflection, observe in pairs or small groups, and take careful notes. Asynchronous observations from the same site can also be a shared foundation for practicing argumentation and responding to readings, and they can be the core of major assignments in different genres that encompass conceptual and practical strategies. These strategies are mappable across different kinds of courses and different student populations. At the center of our ecopedagogy—both local and global—is patient, close attention to place and environment that is in current times both seriously under threat and more necessary than ever.

Acknowledgment

We acknowledge that the territory on which Boston University stands is that of the Wampanoag and the Massachusett People. Our classroom, BU's campus, and local sites such as Hall's Pond are places to honor and respect the history and continued efforts of the Native and Indigenous community leaders which make up Eastern Massachusetts and the surrounding region. This statement is one small step in acknowledging the history that brought us to reside on the land, and to help us seek understanding of our place within that history. Ownership of land is itself a colonial concept; many tribes had seasonal relationships with the land we currently inhabit. Today, Boston is still home to Indigenous peoples, including the Mashpee Wampanoag and Wampanoag Tribe of Gay Head (Aquinnah).

Notes

1 Place narrative assignment provided on the companion website.
2 Materials for these discussions provided on the companion website.
3 Observation handout provided on the companion website.
4 Questions handout provided on the companion website.
5 Activities provided on the companion website.
6 Practical proposal assignment and additional proposal materials provided on the companion website.

References

Coker, J. S., 2017. Pedagogy and place in science education. In D. Shannon and J. Galle, eds. *Interdisciplinary approaches to pedagogy and place-based education*. London: Palgrave Macmillan, 71–83.

Giezentanner, B., and Eunson, D., 1996. *A plan for Hall's Pond Sanctuary*. Lincoln, MA: Massachusetts Audubon Society Environmental Extension Service. Retrieved from http://friendsofhallspond.org/a-plan-for-halls-pond-sanctuary-optimized.pdf.

Hassall, C., 2014. The ecology and biodiversity of urban ponds. *WIREs Water*, 1 (2), 187–206.

Hultgren, J., 2015. Nature, place and the politics of migration. In H. Bauder, C. Matheis, and N. Crook, eds. *Migration policy and practice: Interventions and solutions*. London: Palgrave Macmillan, 127–148.

Matsuda, P. K., and Silva, T., 1999. Cross-cultural composition: Mediated integration of US and international students. *Composition Studies*, 27 (1), 1530.

Shannon, D., and Galle, J., 2017. Where we are: Place, pedagogy, and the outer limits. In D. Shannon and J. Galle, eds. *Interdisciplinary approaches to pedagogy and place-based education*. London: Palgrave Macmillan, 1–8.

Thoreau, H. D., 2000. *Walden and civil disobedience*. Edited by P. Lautner. Boston, MA: Houghton Mifflin.

Van Dooren, T., Kirksey, E., and Münster, U., 2016. Multispecies studies: Cultivating arts of attentiveness. *Environmental Humanities*, 8 (1), 1–23.

6

SAUNTER LIKE MUIR

Eco-Challenges and Experience Projects in Introductory Environmental Ethics

Amanda Hayden

During the American unit of my "Introductory Environmental Ethics" course, students sense right away John Muir is special. With his long "hipster" beard, sweet suit and hat, and crooked walking stick, he makes an impression. When I share that he walked from Indiana to Florida, at least one student pipes up in genuine admiration, "that's hardcore." This opens a perfect place to relate Muir's preference to *saunter* through nature instead of hiking. If we view nature with reverence, like spiritual pilgrims on their way to the Holy Land, should not we *saunter* reverently? Sometimes I demonstrate the distinction in front of the class. For *hike*, I exaggeratedly stomp with wide, skiing arms, from one wall to the other, as purposefully as I can. For *saunter,* I mosey back across with hands in pockets, stopping to "smell" a flower. When it works, amused giggles may ripple across the classroom, and it gets their attention, but I also aim to invite deeper consideration—what does it *look like* when we slow down and saunter *in our lives*?

I invite students to ponder this larger metaphor of being, taking each step in their daily lives carefully, mindfully, with engagement, with reflection. As someone who also teaches about Buddhism, I cannot help making the connection here to the cultivation of mindfulness as daily practice. What better way to manifest this than in our mindful steps to reduce our consumption and waste? To take a closer look at the real consequences of stomping through life and not paying attention to our environment our ourselves within it. The "hiking" in our lives can lead to us wastefully discarding water bottles and coffee cups in our trail, plowing through consumption without giving thought to what's ahead or what will be left behind. I challenge students instead to "be like Muir" during this course, to slow down and examine each step, each choice, each purchase, each discard. To *saunter*, not hike. I don't want them to miss something special.

The course also explores several areas of environmental philosophy, including historical context, religious influences (Eastern, Western, and Indigenous),

DOI: 10.4324/9781003221807-6

minimalism, food ethics, animal rights, social and deep ecology, eco-feminism, and eco-psychology. The final grade is comprised of several assessments, many of which focus on cultivating mindful activism. These include weekly discussion posts and quizzes, attendance and participation, a midterm and a final exam with objective and written portions, and, finally, the Experience Project.

The Experience Project is a high-stakes assignment, and a substantial influence on the final grade; this chapter details student work and feedback for it. Additionally, I offer a few extra credit *eco-challenges*, which can serve as a replacement for a missed discussion post or quiz for students. While they are lower stakes and optional assignments, this chapter demonstrates their potential for much bigger impact. Both assessments are anchored in the philosophy that incremental change can start at an individual level, with many small steps resulting in much larger cumulative outcomes, a ripple effect that continues far beyond one small step or choice.

Both assignments are also structured in a way where students essentially may design their own learning platform and determine most of the criteria for their assessment. As I note later in the chapter, I often refer to this as "choose your own adventure" and in the two decades I have utilized it, it has consistently resulted in extremely high completion rates, grade averages and importantly, student satisfaction and positive feedback. In other words, the approach works.

Mindfulness is the binding thread of all of this—paying attention, truly listening, *really* looking closely, examining, knowing when to be silent and observe—within the assignments and the course itself. With pedagogy that holds immersion learning as direct, effective, and successful, students can directly and confidently apply their knowledge, making the connections for themselves. In the context of this environmental ethics course, it gives them the opportunity to see themselves as part of more-than-human nature, and this, hopefully, impacts not only their perspective but also their actions.

Experience Project

> This project was a joy to create. I felt stretched artistically and … it left me feeling … hopeful.

The Experience Project is a required project with significantly detailed objectives, a variety of options, and the goal of having a positive impact on learning application, retention, and analysis. The parameters allow students to dive deeper into material, tailoring it to their own interests and questions. It demands significant student commitment and engagement, both in quantity (4–5 pages) and quality.

Students have several options for their "experience," including service-learning (volunteering), keeping a "Recycling Photo Diary," visiting a local art museum, or what has become known as "Door Number 4," a unique creative project beyond the three main choices. Students use this opportunity to apply what they have learned from the module lessons and material into "real" life—their own lives.

FIGURE 6.1 Senses Hike, from Experience Project.
Source: Ash Moore.

Recycling Photo Diary

> It put it right in front of my face—and it was not pretty.

Though recycling is emphasized as a last resort and not the "go to" for sustainability, there is legitimacy of its inclusion, since, for many students, it may be a gateway to other eco-actions. This option requires students to collect as many recyclables as possible for any three-week period during the semester and photo-document the pileup. With guidelines on storing and sanitation, I instruct them to keep the collection (ideally) in visual sight, in open bins, in a kitchen, mudroom, or walkway, so the growing accumulation confronts them daily.

Students can be drawn to the project initially due to the lesser quantity of writing required compared to other options. However, they quickly realize that nearly a month of consistent collection, containment, and thorough documentation and analysis is a commitment, and, further, quite inconvenient. My guidelines state, "This *will* be an inconvenience. It will get in your way. It will be tempting to just throw things away, so you don't have to see them." I reassure them there is a point of daily visibility, to see how much it all adds up, every day, in front of them. Students have freedom in how they execute the visual aspects of the assignment. Often during the progression of photos, they will include their cat, roommate, or sibling as a size comparison to the pile. Some students like to include numbers of how many tea bottles or frozen dinners they've consumed or detailed charts to accompany their photos. The requirement of photo-documenting the growing "pile-up" is the foundation of the assignment, but students can create their own thoughtful narrative along the way.

I also provide guidelines for written documentation of the experience. Student must document the "pile-up" at least every two to three days and provide ongoing reflections on the steady accumulation as well as a final summary. Prompts for the reflective component include:

- What do you see on Day 2? Day 4? Day 15, etc.?
- What is taking up the most space? Plastic? Cardboard?
- Do you have more plastic than paper and, if so, approximately how much more?
- What frustrations have you had with collecting every little thing?
- How does your trash can look in comparison? Is it less full than usual?
- What do your parents/roommate think of this project?

At the end of the three weeks, I encourage students to take the collection for recycling, writing a summary to end their documentation. They may share the difficulty of the assignment, how often they wanted to give up, and whether they found it to be easy, challenging, or somewhere in between. They often reflect honestly on whether they may try to change any habits and how it affected the dynamics of the household. A bonus is when a student's roommate or family get involved in the project, which is not uncommon. Overall, they comment on facing their consumption first-hand:

- "I did *not* realize how much I order takeout."
- "I really didn't think I'd have much, but my bins were legit overflowing by the end."
- "Wow, my boyfriend and I drink a *lot* of soda."

I feel it is important (and kind) to reassure students the goal is not to shame. I reiterate this in the written guidelines, verbal expression in class ("this isn't about guilt but about becoming more aware"), and in individual written feedback, if needed. The emphasis on awareness is something most often reflected in student summaries:

> The biggest habit that this project made me change was the mindlessness of just tossing everything into the trash can. It made me stop and think about what I was doing and the importance of making small changes. It also made me realize that almost every product we purchase comes in packaging, whether plastic or paper.

Adaptations

Flexibility is key. Due to the inherent variety of choices on the part of the student, I find obstacles to completion rare. If issues such as cost, access, mobility, or inclusivity do arise, my students and I work together to come up with a solution that works well in this situation. Regarding the issue of cost, I do not require any options that charge a fee. Sometimes students will ask if they can participate in something on their own that may have an admission cost, but that is ultimately their choice and is never required.

Regarding the recycling photo diary, what if a student is already a self-described "avid recycler"? One student of mine decided he would benefit more from completing the assignment at his grandmother's house. During the three weeks, he purchased for her reusable paper towels and sustainable cleaning products, and he encouraged her to switch from her daily use of paper plates:

> At first, my grandmother laughed at me when I stopped her from throwing things away. But, by the end of a couple of weeks, I calculated we saved approximately 231 paper plates and 5 rolls of paper towels! It led to (real conversations) with her about other environmental issues and (I really hope) this will change, at least some, of how she handles her plastic and waste.

The environmental art option is another solid choice, where students seek out any category of artwork relating directly to our subject matter (i.e., nature/landscape, sustainable, repurposed art) and then may choose to focus on one piece in detail or several pieces within that theme. The summary guidelines prioritize what students see, feel, and think the artist is expressing and their reflection on how the art relates to our course subjects. Typically, students physically visit a museum or installation; however, if physical access is limited, I provide a list of virtual exhibits and museums, which give students the opportunity to visit art exhibits all around the world. The happy accident of this adaptation is that it has led to some unique and extraordinary finds.

The *Ghost of Consumption*, by Yinan Lui (2019), for example, is a sculpture made of repurposed pieces of refuse. A student searching for "recycled art" found this piece, admitting at first that she, "just saw it as a bunch of trash covered in plaster." However, once she took some time to really look at it in detail, she, "realized it represented the ghost of these items, how they linger with no place to go…that landfills are essentially graveyards for our waste, how these figurative ghosts come back to haunt us."

With mobility and access issues, adaptation is necessary. We saw this first-hand during the shutdown. How does one complete an immersion project if you are in quarantine or unable to travel? Easily and effectively, as it turns out. The "National Park Virtual Visit" has become one such adaptation. I provide a list of websites for several National Parks giving students the option of "touring" one of these beautiful places, descriptively reflecting on their experience. One student reflected:

FIGURE 6.2 Students viewing environmental themed art at Dayton Art Institute.
Source: Amanda Hayden.

> During the virtual dive at Dry Tortugas National Park, it immediately made me think of the Great Pacific Garbage Patch plastic bag video[1] we watched in class. At Kenai Fjords, I went into a crevasse (which was terrifying—and I wasn't even actually there! I had to remind myself of this to calm down) and was able to get up close to a large glacier and even listen to it under water! Who knew there was such thing as a melt water lagoon? Not me.

Certainly, no one is proposing it is the same experience as visiting in person; however, this option became immensely popular when typical options weren't, with many reflections like this one, students enthusiastically "experiencing" a place they had never been to, often vowing to go see it in person. Even in non-pandemic circumstances, this virtual tour is a worthy option to keep for those with severe mobility or accessibility issues.

Finally, sometimes a particularly ambitious or creative student will ask to do something "outside of the box." With my assurance this choice will likely involve more work and time than the other options, eager students accept the challenge, consistently producing extremely insightful, creative, impressionable work, work I often utilize to inspire future classes. With this, "Door Number 4" was born.

As part of his Honors component, an avid cyclist proposed to trek a loop across three states, documenting the trip through his newly honed lens of environmental awareness. We discussed possible backup plans in case of weather or other challenges, but he followed through with this ambitious undertaking, completing the trip. He provided a photo-documented project of each day and written reflections of parks he rode through, pollution and litter he encountered, stark contrasts between industrial areas and rural areas, swerving to avoid sprays of agricultural fields, and his overall reflection of how differently the world is from a bicycle. He gave a presentation to the class and the campus newspaper featured an interview with him about his adventure. In this article (Stanger 2019), he expressed it was learning about Muir walking from Indiana to Florida that inspired the whole project in the first place.

When everything shut down during the COVID-19 pandemic and classes were taught entirely remotely, I set my bar realistically for our assignments; we were all just trying to get through the semester. Nevertheless, one remote student created and submitted a vibrant, colorful, and thoughtful 10+ page project. Due to us not being face-to-face, and aiming for optimum sustainability, she completed the entire project through her tablet. In an engaging scrapbook style, she detailed her interactions as a metro-park educator, documenting programs, weather changes, moon cycles, food ethic reflections, the cicadas' 17-year return, worm bins, conservation efforts, keeping chickens, wildflower education, patio gardening, rehabilitation of wild animals, meditation, and more. From beginning to end, the project is a clear expression of who she is as a student and as a human being and a joy to read.

By exercising flexibility in options, not only do students benefit, but as the instructor, I am given an incredible opportunity to witness passionate, impactful, experiential learning. "Door Number 4" opens the way for numerous expressions like these that would likely not have happened otherwise. By embracing a student's creative lean away from a "one size fits all" assignment, they can find a perfect fit for them.

Eco-Challenges

> I am really shocked at how all of these little things add up.

As a response to the Muir "saunter vs. hike" discussion, I invited students to spend some time "sauntering" in nature, whatever that means to them. It could be as simple as walking to a local park, sitting under a backyard tree, or relaxing near a creek or fountain. The parameters were purposely loose, with encouragement to keep Muir's preference to "saunter" in mind. Afterwards, students wrote a 1–2-page reflection on this experience. Not surprisingly, students often concluded that participating in this exercise resulted in feeling more peaceful, content, and at ease. Some shared memories of camping trips as a kid or school field trips they enjoyed, while others simply noted feeling calm during a stressful time of the semester, usually some version of "I could focus on what's important."

After seeing how positively students responded to this, I thought to myself, "what if I offer this for other modules as well?" I developed Eco-Challenges as a result. In contrast to the Experience Project, Eco-Challenges are simple, low-stakes, optional, extra credit "mini-immersions" that invite students to delve deeper into the material (and provide a little extra grade insurance). Each challenge relates specifically to that week's module. "Go Local!" deepens student understanding of bioregionalism by looking for local products, viewing labels, and investigating where their purchases are traveling from. During our Animal Rights unit, students can "Tell your Rescue Story," describing a rescue pet who has inspired them. This is, hands down, one of the most completed options and, further, one where students really engage and open up. "No Impact You" (inspired by the film *No Impact Man)* asks students to participate in minimalist actions (reducing energy use, decluttering, micro-changes). "Connect by Disconnecting" challenges them to take a truthful look in the mirror (screen) about our tech dependence with options to reduce screen time, put away phones during conversations, and set timers for social media.

Water Bottle Challenge

> This "little challenge" opened my eyes to a *huge* issue I was contributing to.

This challenge focuses on one of the main culprits of plastic consumption—single-use water bottles. Many resort to plastic bottles for convenience on the run while dropping kids to school, commuting, or after school/work activities, and this challenge gives students the opportunity to be mindful about this disposable go-to. Students receive a "did you know" summary of plastic bottle stats, including the amount of carbon dioxide used; how it can take multiple bottles of water, energy wise, to make just one; how even though plastic bottles are easily recyclable, the vast majority end up in landfills; and, of course, annual cost.

Students track their plastic bottle consumption over a few days' time. Two to three days is the minimum suggestion, but some choose to take on the challenge for the week. They record where they purchase their drinks, how many bottles they accumulate, and what changes they might make during this time to reduce (such as using a reusable bottle instead). I encourage them to reflect on what limiting this consumption looks like for them and how it affects their daily routine. Prompts, such as, "Was it inconvenient? Did it save money? Do you feel more aware of your plastic accumulation?" get students thinking about the broader implications of the experience in their daily lives.

This challenge is one of the easiest to execute due to its universality (everyone drinks water!) and flexibility. A given is that students are coming into the course at all different phases. Some are aware of the issues but are not yet committed to reduce consumption, while others have already made some changes but desire more consistency. For some, any change is uncharted territory and an improvement, after initially voicing a preference for bottled water for many reasons, from taste to convenience. The experience can be quite enlightening for students:

> The first day, I counted a whopping thirteen bottles of water … in one normal day. In class, I learned energy-wise, this is the equivalent to thirty-nine bottles of water. I buy a 40-pack case of water and so, basically, energy-wise, I am consuming almost an entire case of bottles every single day! And this doesn't even consider how much the rest of my family consumes!

Some students ask, "I already use a reusable water bottle, can I still do this challenge?" *Yes,* I tell them, with the urge to focus on consistency. I encourage the challenge to be a wake-up call to focus on a wider circle of plastic use (coffee, soda, energy drinks). One recent student (and mother of four) declared, "Coffee is my best friend! I even have a shirt that says, 'Half-Human, Half-Coffee!'" She knew immediately her focus would be her habit of hitting the drive through for a pricey latte when dropping her kids to school:

> What started as a couple of times here and there as a treat had turned into every … single … day. For the challenge, I forced myself to buy coffee grounds at the grocery store and make coffee each day at home. I even got a Starbucks mug (just to make it "feel" like it usually does). I did this every day! I got through the week and even did it through the weekend! Monday came, the challenge was over, and so I pulled into the coffee shop line as my reward. Then something just went off in my head, "You did this for a whole week. You're in a line with 9 cars ahead of you" so I got out of line. I dropped my kids off to school, came home and made coffee. I want my children and their children to have a healthy planet. It starts with me.

Students express feeling empowered by immediate benefits, including increased water intake, time saved not waiting in lines, and an overall feeling of being part of "the solution" and not "the problem." At the community college where I teach, a good portion of my students are parents and/or working multiple jobs on a tight budget. Understandably, it's often the cost factor which is the major eye-opener:

> I did the calculations. To financially support my plastic bottle consumption, I pay $256.65 yearly. And that is only if I buy my water in bulk, and this is just for myself! If I factored in what the other four people in my family consume, [it is nearly] $800 a year!

Whether students start with water bottles, choose to limit their soda/caffeine purchasing, or start making coffee at home instead of driving through the line, students find their own way to take on this challenge with the same goal—reduce single-use plastic consumption. One of the most fulfilling aspects of the eco-challenge summaries is how many students end up involving their friends, family, and even co-workers. It is not a stretch to say that what begins as a "little" challenge has the potential to have quite a significant impact on students, both in the present and long term.

Adaptations

I design Eco-Challenges for students to choose at their convenience and to be as accessible as possible: free, easy to get to, local. I encourage students to adapt the challenge to whatever environment they are already in, which is important for students with limited access, mobility, or transportation. Offering a variety of choice is key.

As many of my students work long hours outside of school, finding time to complete a challenge at home can be daunting. Occasionally students apply the challenge in their workspace instead. After noting the abundance of Styrofoam her workplace used in shipping, one of my students decided to take on the "4 Rs" (reduce, reuse, recycle, refuse). She collected dozens of containers headed for the landfill, and, after a dive down a Pinterest rabbit hole, her family joined her in "Reuse," transforming them into vibrant, functional planters for their porch.

"Go Local!" gives options of visiting the grocery store with a "label magnifying glass," visiting a metro-park close to home or work or perusing a neighborhood farmer's market. If none of these options are feasible, I encourage students to look around their neighborhood or apartment building, noticing any trash, litter, or recycling options, garden/compost availability, and green space (or lack thereof).

While there are parameters and specific suggestions, students are not required to do anything when it comes to their specific choice. Out of a wide breadth of options, students ultimately choose for themselves how far they want to go in applying the challenge and what best fits both their interests and needs. This "choose your own adventure" approach, both with Eco-Challenges and the Experience Project, can be excellent teaching insurance. Since the students choose their own topics and ways to be assessed, ultimately there is an extremely low incompletion (and low complaint) level. On the contrary, students often follow through in ways that reach far beyond the assignment requirements.

Analysis and Assessment

> This assignment has truly inspired me to be a better role model to my children.

As part of research outcomes regarding global citizenship and cultural diversity for a department review, I collected data on the Experience Project as a common assessment tool across several different courses in both the humanities and religious studies over a period of eight semesters from 2013 to 2017. I focused on writing skills, examination of personal values, critical thinking, and appreciation of diverse perspectives. A sampling of 377 students' experience projects over eight semesters led me to find an 84 percent completion rate and an 87 percent grade average. Further, there was an extremely low plagiarism rate due to its innate reflective and personal nature. Overall, students navigate the project with high quality and authenticity and commonly go beyond the minimum requirements. Students overwhelmingly cite this project as a favorite aspect of the course in their written feedback.

Because of its inherent adaptability to be tailored to the specifics of course topics, the Experience Project can be used across disciplines beyond environmental ethics. In fact, I had been using it in my religion courses for over a decade before I integrated it into the environmental classes. In addition to art and service learning, my religion students have the option to respectfully visit and observe an unfamiliar religious service (mosque, synagogue, church, temple, gurdwara). A department request for a universal assessment for cultural diversity and global citizenship outcomes led me to share this project with faculty in several other disciplines (history, modern languages, political science). My colleagues then created their own adaptations by applying choices such as observance, art, and volunteering to their specific discipline(s).

Regarding the eco-challenges, when an assignment is labeled "extra credit," no matter the content, students tend to take them on with a positive mindset and enthusiasm. Instructors may choose any number of ways to grade extra credit, with some giving a simple pass/fail, while others may have a more complicated and detailed rubric. Since they are worth up to 10 points in my course, my criteria are that students do the following:

- Participate authentically in the challenge and consider what it asks (i.e., did they make any attempt at changing habits, and if so, how?).
- Reflect on the challenge and how it applied the class material (i.e., did they make a connection with the "theory" and the application in their own life?).
- Complete a 1–2-page reflective summary that shows they completed the challenge with honesty and integrity.

Any of these eco-challenges have the possibility of being offered as a bigger assignment, with more depth and analysis. The potential for this is apparent, as a few of my students have utilized their Eco-Challenge experience as a gateway or springboard into their larger Experience Project, exploring the challenge in more depth and with more time. For example, a recent student had already completed the "Disconnect to Connect" Eco-Challenge by reducing her social media use over a few days at home. While planning a backpacking trip for the larger Experience Project, she decided to implement what she gained from the Eco-Challenge, combining her mindfulness of minimized social media use into her weekend camping trip. She reported in the project summary that by being motivated to stay off her phone and not check Instagram stories or scroll, she felt legitimately more present in the woods, increasing her quality of her experience.

After completing the challenges, students consistently conclude with a desire for more accountability, reporting that while they were aware of the issues before the class, the Eco-Challenges, specifically, gave them the push to start making some changes.

Conclusion

> I considered myself an advocate for social justice and environmentalism before, but this class put it in perspective, implementing what I've learned into my work and home life.

Eco-Challenges began as a response to the giggles of an improvisational classroom exercise, yet the consistent, positive results became a springboard for a larger library of experience-based methods in this introductory environmental ethics course. Likewise, what began as a simple "go experience this" immersion project has, over time, been seamlessly woven into several courses, consistently cited by students in evaluations as one of their favorite parts of taking the course. I have graded thousands of these challenges and projects over the years. One College Credit Plus student (high school student taking college credits) did something quite extraordinary with both.

Excelling in the course (as many of my CCP students do), this student certainly did not need the points for her grade. She chose to complete multiple Eco-Challenges, she told me, for "the experience of it." She reached out to me the following semester, indicating that the Eco-Challenges inspired a desire to follow up with similar themes for her current Honors Society project. We met, batted around some ideas, and I mentioned that several faculty had long hoped our cafeteria, serving thousands of students a week, would switch from Styrofoam and plastic to-go containers to more eco-friendly products.

A month later, I walked into our cafeteria and stopped in my tracks. Sustainable paper boats and compostable containers were now neatly stacked where Styrofoam and plastic used to overflow. I thought to myself, "this must be a coincidence." It was not. A few days later, the student sent me her Honors draft, and I read with amazement as she described her meeting with the cafeteria director and her gentle, but firm, push for this much needed change, and the positive reception that followed. I teach and talk about these issues every week, but this was manifestation—actual results! This local 17-year-old high school student used her voice to directly affect the entire cafeteria's carbon footprint. And drumroll … she insisted the Eco-Challenges—those small, optional assignments that she did not even need to complete—served as the inspiration. Does it get better than this for an instructor?

My hope is readers of this chapter consider the nearly limitless ways they may seamlessly integrate experiential learning into their courses, not just with environmental philosophy, but with virtually any class subject. Immersion learning empowers students to be part of the solution and empowers instructors to witness students making connections themselves through implementation of course material into tangible life. It is hopefully why we teach these subjects in the first place: to engage students, to invite them to confront broader issues directly, and, ultimately, to connect with each other, and with themselves. Experiential learning starts with encouraging students to do one seemingly small thing: to *saunter*, and not hike, through this beautiful, unpredictable, and rugged life.

Acknowledgment

We acknowledge in the Ohio Valley, we reside on the original lands of the Shawnee, Leni Lenape/Delaware, Ottawa, Miami, Wyandot, and Seneca-Cayuga peoples. It is necessary to understand and honor the sacred value this land holds to Indigenous nations.

This is not past tense. These tribes were eventually and forcibly removed from Ohio. We must sit with the discomfort of our historical context and that this colonialism is *ongoing*, not just a mark on the past, continually manifested in environmental devastation, the disproportional rate of missing/murdered Native women, poverty and illness and land/treaty injustice.

We pay our respects, not just by acknowledgment, which is a first step, but by committing ourselves to further education, awareness, and action.

Note

1 The student is referring to *The Majestic Plastic Bag: A Mockumentary* (Heal the Bay 2010).

References

Go Green BK Team, 2019. Eco-friendly art: Upcycled sculpture "The Ghost of Consumption". Retrieved from https://gogreenbk.org/staging/2019/01/eco-friendly-art-upcycled-sculpture-the-ghost-of-consumption (accessed August 2, 2021).

Heal the Bay, 2010. *The majestic plastic bag: A mockumentary*. Video recording. Retrieved from https://youtu.be/GLgh9h2ePYw (accessed August 2, 2021).

Stanger, N., 2019. Sinclair Student Completes 180 Mile Cycling Adventure. *The Clarion*, 19 December. Retrieved from www.sinclairclarion.com/home/tartan-news/2019/12/19/sinclair-student-completes-180-mile-cycling-adventure (accessed August 2, 2021).

7

DECOLONIZING OUTDOOR EDUCATION

Reading Muir in Alaska and Fly Fishing on Lingít Aaní

Kevin Maier

The University of Alaska Southeast's mission—emblazoned on ubiquitous banners and posters on campus—proclaims that "student learning" should be "enhanced" by "the cultures and environments of Southeast Alaska." It is a charge faculty take seriously, routinely bringing our students into the communities and environments around our campus. This is especially true of our humanities-oriented Outdoor Studies Program, one of several interdisciplinary degrees in which my environmentally oriented courses are core requirements. Outdoor Studies students spend 40–50 days in the field each year, honing skills in climbing, skiing, mountaineering, glacial travel, kayaking, and navigation. My training is in literary and cultural studies, so I work mostly on the more traditionally academic side, but, on average, each year I still spend seven nights and upwards of 15 days in the field with students from a wide range of backgrounds and degree programs. Although I wasn't initially convinced I should sacrifice classroom time for field time, after 18 years of taking students outside, I believe these outdoor components do in fact enhance student learning.

Although we have strong institutional support and infrastructure for taking students outside, we still occasionally have to offer arguments for the outings when we put together course schedules. My justifications to administration note these outings are essential responses to an increasingly indoor, screen-oriented, and nature-deprived student body. This is almost always enough to win a weekend spot on the calendar, and to include a modest lab fee to cover the cost of travel, the wages of a student teaching assistant (TA), and evacuation insurance for each participant. If I meet resistance, or if I really want to hammer home the need for these outings or a particular course fee, I argue that my field excursions are specifically designed to strengthen the arguments of my courses (admin prefer that I call these "learning outcomes"); which is to say, I use outings to amplify environmental justice critiques of traditional environmental politics. In theory and in practice,

DOI: 10.4324/9781003221807-7

then, I frequently use outings as an opportunity to think critically about the Western outdoor-education model that champions outdoor experiences as vehicles that move students toward environmentalist positions.

By "think critically," I don't mean to dismiss the benefits traditionally understood to emerge from outdoor education. There are countless reasons to have a polis more attentive to the ecosystems that sustain us. To ensure my course excursions increase awareness, I embed ecological literacy into all the outings: I somewhat jokingly assign "Big Year"-style bird identification contests; at trailheads, I have students present on weather predictions and potential risks in our travel (Avalanches? Hypothermia? Bears?); others present on management regimes determining the rules of the land we will move through (Is hunting allowed? Will we encounter motorized vehicles?). All of this is premised on the conventional idea that better awareness of what's beyond the pavement will inspire students to be better stewards. Being better stewards requires attending to the historical and political dimensions of these forays afield, too, and this is where the outings have the highest impact.

In what follows, I describe two experiential education trips to discuss my efforts to explore this history and to, at least in some small sense, "decolonize" outdoor education: first, I detail an overnight trip to a backcountry public-use cabin to discuss John Muir's *Travels in Alaska* as part of my 300-level literature and environment course. Second, I discuss a week-long expedition to carry out a service-learning project and fly fish for steelhead that concludes my 300-level humanities class called "Salmon, Sport, and Society."

Before moving to more practical observations about these course outings, as a non-Indigenous scholar, I want to pause here to note that I'm invoking the word decolonization with caution. A contested term marking myriad efforts to push back against the impacts of colonialism, the term decolonization refers to more than simple political processes. As Linda Tuhiwai Smith (1999, p. 98) puts it, "decolonization, once viewed as the formal process of handing over the instruments of government, is now recognized as a long-term process involving the bureaucratic, cultural, linguistic and psychological divesting of colonial power." A vibrant body of scholarship has emerged over the last two decades, engaging important debates about how the term plays out in research and pedagogy. In my discipline of environmental literary and cultural studies, Kyle Powys Whyte has most clearly described the intersections of colonialism and environmental justice. In a large body of work, Whyte (2018a, p. 125) reminds us that conventional environmentalist positions like wilderness preservation reflect settler worldviews, which strategically "undermine Indigenous peoples' social resilience." Pointing out that settler privilege is persistent in environmental thinking, Whyte (2018b) nevertheless asks non-Indigenous allies like me to "be honest about decolonization."

In their important essay "Decolonization is Not a Metaphor," Eve Tuck and K. Wayne Yang similarly note that as the term finds currency in the academy, it risks becoming a buzzword, standing in for too wide a swath of social justice concerns (Tuck and Yang 2012). Tuck and Yang also worry that the too-easy adoption of the term can mean it is deployed to assuage settler guilt rather than effect real change. They enumerate these latter rhetorical moves, appropriately calling them "settler moves to innocence." As they put it (ibid., p. 10), "settler moves to innocence are those strategies or positionings that attempt to relieve the settler feelings of guilt or responsibility without giving up land or power or privilege." Tuck and Yang posit "an ethic of incommensurability" as one alternative, suggesting allies and advocates strive to "unsettle innocence." For Tuck and Yang, this unsettling means noting how efforts to create solidarity among disparate justice concerns might not only be detrimental to broader decolonization efforts, but that they in fact might perpetuate settler ideologies.

There's a lot more to say about the intersections of environmentalism and settler colonialism, but for the purposes of this paper, the important thing to note is that my tentative efforts to decolonize my pedagogy hinge on course outings that show students that sometimes the act of going outside to experience the "environments of southeast Alaska" in ways we've been told are not only innocuous and fun, but educational and good, can in fact be detrimental to Indigenous self-determination. To unsettle this innocence, then, I orchestrate course outings around this "ethic of incommensurability." In fact, if I were to posit a single "learning outcome" for my outings, it would be for my students to realize that things they like to do (hiking, skiing, camping, hunting, fishing) can be used to leverage heightened ecological awareness, providing a pathway to an environmentalist politics, but they are often also incommensurate with a commitment to social justice that shapes their identity.

This outcome is especially important at UAS, not just because our mission declares it so, but because our institution sits in the heart of *Lingít Aaní*, Tlingit ancestral land. We have committed significant resources to Tlingit language revitalization, and our programs have begun the difficult work of acknowledging and addressing the violent history of colonialism and its contemporary expressions. We have a long way to go, and there is shocking resistance in many of the institution's structures and in sometimes-surprising corners. I want to turn to two examples of on-the-ground pedagogy that represent my attempts to leverage experiential education in service of a broader effort toward racial healing.

English 303: Literature and Environment

My semester-long classroom-based literature and environment class always includes a weekend outing where we walk several miles to a rustic Forest Service cabin for two days of discussions. In almost two decades of teaching this survey class, I've maintained a focus on so-called classic American texts—*Walden, Sand County Almanac*, and *Silent Spring* are mainstays—while also shifting to consider more local concerns and, lately, the imaginative responses to ecological realities of our Anthropocene moment. The classic texts represent good opportunities to establish the historical origins of our current ideas, while also elucidating the social justice blind spots of the American environmental movement—for all his useful narratives about living simply and attentively, Thoreau rarely mentions the Indigenous inhabitants of Walden Pond, choosing instead to figure himself as more Indigenous to the land than most, for example. The more contemporary texts, like Octavia Butler's *Parable of the Sower*, Leslie Marmon Silko's *Ceremony*, or Richard Powers's *Overstory*, show that intersectional issues are more important than ever as we think about the ways environmental concerns connect to social issues. Sometimes the vagaries of course scheduling mean the course outing happens in the first few weeks of the semester, providing key community forming moments for the class; sometimes it falls toward the end of the semester, and we wrap up the course with final presentations. The last several years I've been lucky enough to get a mid-semester outing, and I've melded the chronological march through American environmental ideas such that we can read selections from John Muir's *Travels in Alaska*. To really amplify course learning outcomes, I have been especially careful to shape the dialogue around this classic nineteenth-century text to highlight the intersections of colonial ideology, science, and Indigenous ways of knowing that we can find in the text.

To accomplish this, students read selections from the book in advance of the trip, and we spend at least one classroom session discussing Muir's formula of first-hand experience in the mountains leads to a politics of preservation. After this discussion, usually enhanced by student presentations on Muir's life, the critical reception of the book, and some relevant history, we spend almost all of our next bi-weekly 90-minute class sessions on trip logistics. Having distributed a detailed packing list the week before, I have a more capable student bring a pack to class to model appropriate overnight gear and clothing; we check out stoves and packs and sleeping pads as needed from our recreation center; and I assign cook teams for menu planning. Mindful of the cost of outdoor gear, I urge students not to buy anything for the trip, to think of our community as a gear library instead. I insist that if you have appropriate footwear and a good attitude, the rest of things will fall into place. It's fun to watch little exchanges happen in class, as students swap rain pants for headlamps for the weekend. Because our Outdoor Studies program courses routinely feature skiing in avalanche terrain, climbing significant summits like Denali, and spending hours each week in high-consequence environments, UAS has necessarily robust risk management protocols. On our logistics day, then, students review and sign waivers, privately share written pertinent medical histories with me, and we make sure everyone is ready to go emotionally, physically, and academically.[1] For some of my students, this is one more weekend outing, and with the destination a heated, hard-sided structure 3–4 miles down a marked trail, it represents a break from navigating off trail to sleep in a tent in the rain on, say, a glacier much further afield. For others, this is the most adventurous and physically demanding thing they have ever done. There is value for having both the experienced, many of whom are hoping for work in the guiding industry, and the less experienced, many of whom have grown up near these trailheads but never stepped beyond the gravel parking lot. The former practices the necessary social and soft skills, and the latter group gets exposure to an amazing and affordable resource in our community.

Mindful of the ableist biases in outdoor recreation and the increasingly fragile mental health of my students, I also make sure there are alternatives to the outing for those who can't attend. This is a good safety valve for students early in the semester, but more "I can't make it" students always seem to emerge at the eleventh hour, so I try to strike a balance between being very encouraging ("We'd love to have you; students almost always say the outing is a highlight of the trip," etc.) and understanding ("don't worry; please propose a short assignment that you think would be the equivalent of the class sessions from the outing and we'll recap the discussions in class next week.")

Our Saturday morning rendezvous before the outing offers one more opportunity to check our gear and preparedness, and, as we load into vans to drive to the trailhead, students prepare us for the physical hazards and challenges we will soon encounter. I've tried to use trail discussion prompts, asking students to discuss course texts or ideas as we walk in, but I've found that unstructured informal exchanges enable students to connect better, so I leave the intellectual walking-work to the birds, simply asking students to pay attention to their surroundings. Depending on the group[2] and the conditions, we walk for two to four hours; moving through Southeast Alaska backcountry almost always requires negotiating puddles, mud, and muskeg, but some years there is deep snow, some years we need ice-cleats, and sometimes the pelting rain slows us down. No matter how long or miserable, as soon as we arrive at the cabin and unload our heavy packs, I pass out candy bars to reward the best bird identifiers and tell everyone to prepare to talk more about Muir's sublime feats in the mountains. Before our discussion can begin, I ask students to find a warm and comfortable place with enough light to read copies of the introduction to a collection of Tlingit place-name maps, Thomas Thornton's (2012) *Haa Léelk'w Hás Aaní Saax'ú / Our Grandparents' Names on the Land.*

Huddled around a hissing oil stove in a poorly lit cabin, and having just read about Indigenous place-naming conventions, our discussions of chapters in Muir's book like *The Discovery of Glacier Bay* take a notably different tone than when we are in the classroom. It doesn't take students long to realize that the landscape they just traveled through, although managed for multiple use by the US Forest Service and surrounded by mountains bearing names like Muir's, is contested space, and the well-maintained trail, the clean cabin, and the relative ease of travel belies a recent history of dispossession. Comparing Western place names (often generic descriptors or personal nouns, like Glacier Bay or the Muir Glacier) with Tlingit place names (which are often process-oriented and imply a relation to a particular lived history), Thornton argues that Tlingit "place names not only define the land but work within cultural systems to maintain people's sense of being and belonging in Tlingit Aani" (Thornton 2012, p. xiii). With this context—and the documentation of how Western culture has systematically erased from the land Tlingit place-names—students readily draw connections between Muir's rhetorical project and settler colonial worldviews. Muir travels to Glacier Bay ostensibly to find evidence for nascent theories of landscape formation through glaciation, taking great joy in stupendous feats of climbing and hiking in dangerous terrain to observe firsthand how glaciers are moving rivers of ice. His prose is infectious and compelling, and it is hard not to be moved by the sheer joy Muir takes in his jaunts afield and the glaciology they enable; but having more recently learned the Tlingit names for the same landscapes where Muir plays and studies, students are a bit more circumspect. Consider the Tlingit name for what is now also known as Glacier Bay: *Sit' Eeti Geiyi* translates to "Bay in Place of the Glacier." Thornton reminds us that this name implies the very process of glacial movement that Muir came to study. Looking across the long glacial Fjord we now call Stephens Passage toward Sit' Eeti Geiyi, students quickly see how Muir's stories of adventure and discovery re-map an already heavily storied and well-understood landscape, facilitating the violent Westernization of Alaska Native peoples and lifeways.

Importantly, empowering students to link a canonical environmental text to historical violence is not done to debunk Muir as a mere racist. Instead, I go to great lengths to show how Muir's glaciology shapes what he sees and how he interacts with the people and landscapes he encounters on his travels, and that the scientific literacy he ushered in has no doubt helped contemporary scientists identify threats to our planet. I'm also careful to ask students how these legacies inform and limit our own interactions with the land. Students identify the incommensurability of Muir's science and Indigenous ways of understanding more readily when we have just moved across the same landscapes we are reading about. These unsettling moments more readily emerge when we are on the land, in the places, and thinking about how history shapes perceptions. They also set-up more traditional classroom exploration that will occur later in the course. While savvy students may be able to see the lacunas in Aldo Leopold's enchanting *Sand County Almanac*, everyone seems especially tuned in to the ways that science sometimes offers too narrow a view and to the problems that arise from a lack of awareness of privilege and power. Indeed, as much as my students love Leopold's twentieth-century update to Thoreau, our field experience helps elucidate how much the contemporary environmental movement is built on a deep exploration of place accessible to very few people.

Humanities 372: Salmon, Sport, And Society

Like the annual course outing in my literature and environment class, the seven-day trip to the village of Yakutat that culminates my spring term Humanities 372: "Salmon, Sport, and Society" course offers more opportunities for these unsettling moments to emerge. The trip is ostensibly to fly fish the internationally famous Situk River for steelhead, and this is both how I advertise the class and why students sign up.[3] However, a semester of classroom sessions prior to the trip prepares students to think critically about the cultural politics of bringing elite sport-fishing worldviews to a predominately Indigenous community supported by traditional subsistence and commercial fishing.

Salmon provide a perfect vehicle for exploring vexing social and environmental problems, especially in Alaska. While we may link salmon to the broader region of the northwest, the bulk of wild salmon that end up on hooks or plates are from Alaska. Increasingly more icon than living entity in the broader region, salmon remain a defining factor in the cultures and day-to-day lives of Alaskans. From Alaska Native peoples who have thrived here for millennia to just-arrived migrant cannery workers, salmon have long sustained people in Alaska. For the most part, humans have sustained wild salmon here as well. Despite the well-rehearsed story of Alaska's salmon management successes, problems are present in today's salmon–people system. Salmon are central to conflicts between rural and urban communities, notably between subsistence, sport, and commercial fisherman. Salmon also swim through tensions over racial justice, and, perhaps most notably, they prove central in discussions about the delicate balancing act between sustainability and the state's need to create revenue through the development of other resources (especially mines that so frequently sit in the headwaters of productive salmon streams.) Moreover, as the rapid changes of anthropogenic climate change create more environmental unpredictability, increasingly complicated salmon and people tensions continue to emerge.

To address these tensions, we spend the semester in the classroom exploring the role sport fishing has and could play in making the salmon-people system in Alaska and beyond more sustainable, equitable, and resilient. More specifically, we consider the potential of the culture of fly-fishing as a response to the tragically familiar story of decline that characterizes the human relationship to salmon throughout their former range. As a humanities course, our concerns are primarily with the cultural and human problems inherent to fisheries management. We nevertheless read widely in the scientific and popular literature of salmon management as well as in the more contemplative angler-written fly-fishing literary tradition, reading popular science texts like *King of Fish* (about which the students host a public discussion) as well as niche essays from anglers like Roderick Haig-Brown, Ted Leeson, and others.

I typically schedule the class to meet every other week for three and a half hours. Part of each class is blocked out for visits from leading fisheries managers, conservation activists, subsistence fisherman, and cultural leaders. Another part of each class session is for honing practical fishing skills (learning to tie flies, cast fly rods, and tie appropriate fishing knots), and a portion of each class session is for preparing a service-learning project. Finally, I ask the students to plan the logistics of the week-long trip, from food to plane tickets and from flies to daily itineraries. It sounds like a lot, but I schedule the breaks between classes and the mix of hands-on and discussion, and practical and theoretical, seems to make each session go quickly.

By the time we arrive in Yakutat, students are set-up to think carefully about how the sport angler's ethos interacts with subsistence users' relationship to the river. Fish, in other words, provide a very good vehicle for considering complicated interdisciplinary problems of our twenty-first-century Anthropocene moment. Our trip to Yakutat, a predominantly Alaska Native village a couple hundred miles up the coast from our campus in Juneau, offers ways to see these tensions playing out in real-time with real communities.

These theoretical concerns are brought to presence when we arrive. Our first stop after disembarking from the 50-minute Alaska Airlines flight is often the fly shop, where we get the fishing report, pick up some last-minute supplies, and the proprietor sells T-shirts that read "I play with my food," with an illustration of a fly-casting angler above the words. Although a joke, poking fun at the catch-and-release sport fishing that draws anglers from around the world, this t-shirt is the first indication that the relationship between visiting angler and local fisherman might be uneasy.

This is amplified when my university students get to the secondary school. The service-learning portion of the course requires the UAS students to spend a week teaching classes to 7th–12th graders as part of their annual "outdoor education" week. UAS students design sessions based on their own particular skills. Over the years, these have ranged from wilderness first aid to iPhone photography, and from basic biology to knot-tying skills. Although the classes are rewarding for instructors and pupils alike, the most striking development is that Yakutat kids who tend to call us "sporties" upon our arrival, lumping us into the category they use to dismiss visiting anglers, know us by name by week's end, and tend to see us more as friends and fellow humans by that time.

It helps that my university students are typically diverse in every sense, reflecting not just UAS's general student demographics of 20 percent Alaska Native and 65 percent women, but also in terms of disciplinary background, as students come from biology, geography, English, outdoor studies, and business. They have varying levels of experience with sport fishing, but often they are already anxious about the ethics of a catch-and-release fishery (Alaskans eat a lot of fish!). The local response further unsettles their sense that sport anglers who practice catch-and-release fishing to save the resource are good conservationists.

Some of the more striking moments in our field course occur when the students host a community dialogue in Yakutat. We do a warm-up community discussion in Juneau, focusing on one of our course texts, so students are comfortable with divvying up the work of bringing voices into the room. The last time I taught the class, we were lucky to have the Yakutat area fisheries biologist from the Alaska Department of Fish and Game, a USFS manager, a commercial fisherman, the mayor, who is also a leader with the local native corporation, the environmental specialist with the Yakutat Tlingit Tribe, and the owner of the local fly shop. Huddled in the Yakutat school library, my students framed questions that emerged from our semester long exploration of salmon, pointedly asking our local experts to comment on the role of sport fishing in creating sustainable and equitable fisheries in Yakutat. Some noted that elders don't want any tourism and would prefer that anglers didn't come from outside to visit; others called it an essential and underdeveloped economic driver. We learned, too, that the commercial fishing industry, once the backbone of the cash economy, no longer provides wages to support families, as the price for their fish is set by fickle global markets.

It was an emotional evening, and my students, who signed up to learn to fly fish for steelhead, learn as much from the community and the environment as they did from books or direct instruction about fishing tactics. They always catch a few fish—never as many as they anticipate, as we all get schooled by what are affectionally known as the "fish of 10,000 casts" for their stubborn lack of interest in taking sport fishing offerings. Instead of simply accepting the sport angler's insistence that a sport fishing ethic of restraint leads to environmentalist politics, the field experience with local youth, Indigenous leaders, and the broader community helps us realize that sometimes sport fishing displaces traditional uses. Students leave a bit more critical of the experiential activist paradigms and more aware of the colonial vestiges of the globe-trotting angler. Students also get to see the spectacular spring migration of birds, breakers rolling in off the Gulf of Alaska on miles-long sandy beaches, and everyone learns the song of ubiquitous ruby-crowned kinglet. The experience, in other words, unsettles the privilege of recreation-oriented engagements with the non-human world while also showing the value of immersive experiences recreation enables.

In both these examples, the experiential component enhances the classroom learning, enabling more sophisticated and grounded inquiry into vexing cultural and environmental problems. In both outings, moreover, the goal is to enhance learning about the limits of conventional environmental thinking without simultaneously debunking or devaluing the positive work conventional environmentalist politics can accomplish. My hope is that this both/and approach empowers students to carry what Tuck and Yang (2012) call an "ethic of incommensurability" into their engagements with not just outdoor recreation, but their day-to-day lives as well. My worry, of course, is that this theorization of the outings marks my complicity in what they call "settler moves to innocence." This tension is one that will continue to animate and haunt all my teaching.

Acknowledgment

I am honored to live and work on land of the Tlingit people, who, since time immemorial, have been and remain stewards of this place. I acknowledge the brutal history of colonization and dispossession that shapes our collective relationship to this land, while doing my best to amplify and lift-up the voices of resilience and cultural revitalization that counter this on-going colonial project. Gunalchéesh.

Notes

1 Our waivers are approved by the statewide Risk Management office. The medical history forms were developed with National Outdoor Leadership School (NOLS) Wilderness Medicine Institute staff to gather information that help Wilderness First Responders (WFR) make better decisions in the field. We also have an Outdoor Studies Risk Management Plan that dictates that backcountry outings require at least one WFR per 8 students. I keep my certification current, and we bring a TA with a certification on most outings. In terms of academic preparedness, students must be passing my class to attend the outing, and if students will miss other classes, I circulate a draft letter that includes a signature line for other faculty showing that they have permission to travel. Increasingly,

evaluating mental health concerns is my most difficult task, and this is an area where I know I need more training.

2 Although capped at 25 students, enrollments have varied from a low of 15 to an overloaded section of 33, but we are almost always too many to sleep comfortably in the reservable cabins designed to sleep 6–8, so several of us bring tents and camp outside the cabin, only coming in to eat and have discussions.

3 As one upper-division elective among many, and a class that comes with an intense schedule, a hefty course fee to cover commercial airfare, and that conflicts with other core requirements on the schedule, I've so far had to advertise this class to get the minimum eight students needed to "make." I can't transport more than twelve students in Yakutat, so the course requires instructor permission and I'm necessarily selective. If I don't already know a student, I ask them to submit a short e-mail letter of introduction indicating how the class fits their needs and why they want to take it.

References

Thornton, T. F., 2012. *Haa léelk'w hás aaní saax'ú: Our grandparents' names on the land.* Juneau, AK: Sealaska Heritage Institute.

Tuck, E., and Yang, K., 2012. Decolonization is not a metaphor. *Decolonization: Indigeneity, Education & Society*, 1 (1), 1–40.

Tuhiwai Smith, L., 1999. *Decolonizing methodologies: Research and Indigenous peoples.* London: Zed Books.

Whyte, K., 2018a. Settler colonialism, ecology, and environmental injustice. *Environment and Society*, 9 (1), 125–144.

Whyte, K., 2018b. White allies, let's be honest about decolonization. Retrieved from www.yesmagazine.org/issue/decolonize/2018/04/03/white-allies-lets-be-honest-about-decolonization (accessed January 3, 2022).

8

NATURE REVISITED

Ecopedagogy in an English–Physical Education Learning Community

Ian MacKenzie and Doug Smyth

"Nature Revisited" is a paired-course learning community co-taught by the authors at Dawson College, in Montreal, Quebec. Ian teaches "Into the Wild: Writing about Nature and Environment," a general education English course that introduces the North American tradition of nature writing by way of classic and contemporary authors. Doug leads an intensive outdoor Physical Education course, "Nature Retreat," which explores skills and mindsets that will allow students to confidently enjoy a range of outdoor activities such as wilderness camping, hiking, and canoeing, as well as flora and fauna identification. Our two courses are enrolled with one cohort of 25 students who learn in traditional classrooms, in urban greenspaces and remote wilderness locations, and on a course blog.

In what follows, we share what we've learned about using learning community curriculum and pedagogy to get students out of the classroom and into the outdoors, including during the COVID semesters of fall 2020 and 2021. We imagine readers of this collection are excited by the possibility of designing learning experiences that are authentic, embodied, experiential, and outdoors, but are also wondering how such an endeavor is practically possible and what ecopedagogical principles might orient a successful effort. Multiple challenges come with co-creating and co-teaching interdisciplinary paired courses. By speaking here of six principles that have guided our efforts, and which instructors can easily adapt to different contexts and purposes, we hope to encourage readers to embark on their own collaborative projects.

Work Together to Get Outdoors Together

First, let's acknowledge that within the North American notion of getting back to nature is a widely recognized strain of rugged individualism. While this mindset is of interest as a *topos* of nature writing and environmental thought, we think it best

DOI: 10.4324/9781003221807-8

to highlight right away that it is not sufficient for getting teachers and dozens of college students safely outdoors together. In fact, we believe that outdoor self-reliance and the reflective mode of environmental thinking and writing are best fostered within a context of supportive relationships. We found three types of "together" to be critical for getting college students outside and learning together.

The first key relationship involved connecting with a colleague who shared an innovative vision for outdoor learning and possessed a complementary professional skill set. On his side, Ian was looking for a collaborator with expertise and experience in outdoor education, someone who shared his love for the outdoors and had an interest in combining the experiential learning specific to outdoor education with the bookish learning at the heart of his literature courses. Ian is a lifetime skier, hiker, and canoeist, but he has no formal certification in wilderness first aid, for example, or in qualifications to teach rock climbing. In short, he needed to team up with an outdoor education specialist with experience in trip planning, budgeting, and logistics; risk management; and outdoor education design and instruction.

Doug's search for a collaborator arose of a desire to address environmental concerns in greater depth than was feasible in his conventional outdoor education intensives. For example, Doug felt that "leave no trace" (LNT) camping principles could be taught as an introduction to environmental ethics. The seven LNT principles provided a basis for making students aware of the impact their behavior might have on trails, campsites, and waterways—and why they should care. His course ponderation, however, allowed almost no time for readings or discussions, and he doubted that students were developing any lasting ethical awareness.

Our first informal conversations marked a turning point. Linking courses would increase Ian's outdoor time with students, and an interdisciplinary approach integrating outdoor experiential learning with analysis of literary texts might address Doug's depth deficit. But a second level of collaboration was essential for "Nature Revisited" to evolve from vision to reality. We needed to forge institutional relationships leading to the approval, staffing, and scheduling of linked, team-taught English and Physical Education courses—definitely not a commonplace of interdepartmental cooperation, at least at our college. Happily, our initial scheme had the good fortune of intersecting with the expansion of Dawson College's learning community program.

Learning communities (LCs) have been adopted by hundreds of North American colleges and universities that are looking for evidence-based approaches that enhance student engagement and learning (Smith et al. 2004). Typically, a single cohort is enrolled in two or more courses in complementary disciplines, and the team-taught pair or cluster is designed around a complex interdisciplinary theme. This also entails numerous administrative tasks, typically involving the registrar's office, scheduling, advising, department chairs and LC coordinators. Participating faculty must adjust to a new level of cooperation and communication, beyond the customs of their home departments. At Dawson there are now well-defined protocols that guide the interactions of teachers, administrators, and staff, but teachers must still be armed with patience and a collaborative spirit.

The payoff for patience is a new world of possibilities for outdoor learning. We pair an intensive 30-hour outdoor education course (six classroom meetings of 2 hours and three day-long/6-hour excursions, spread over 9 weeks) with a standard 60-hour English course (two 2-hour classroom meetings a week over fifteen weeks). We are blocked to co-teach in each other's classes, and one of the weekly English classes sits in a back-to-back position with the Physical Education class on the same day, giving us an entire four hours to get outside. Regular outdoor activities within the weekly academic schedule are thus feasible.

Finally, learning communities pedagogy encourages a qualitatively different sort of relationship between student peers and between students and teachers. This is the third type of "together" that interested faculty must be prepared for. Paired courses mean more intense relationships with students and co-teaching colleagues, a reality to which not all teachers are predisposed. For students, research confirms the payoffs of cohort-based social relationships. Studies suggest that closer student–student and student–faculty relationships in LC cohorts lead to greater time-on-task commitment to learning, and, consequently, to improved outcomes in persistence and achievement (Zhao and Kuh 2004; Kuh 2008). We found these positive outcomes to be apparent from our very first iteration of "Nature Revisited," and they have only increased as our experience and confidence with the LC approach evolves. Our cohorts grow closer not merely by spending more time together but by accompanying each other in tasks that are physically and intellectually challenging. We see students rally together when there is a canoe to load or a tent to set up, and in parallel we see a heightened commitment to support each other in the learning activities and formal assessments of our courses. Friendships are formed more easily in an environment where collaboration is valued over competition, as can be the case in traditional classrooms. One student noted, "Because I felt very comfortable with the other students, it was easier to participate and engage in the classroom and also it increased my self-esteem by being in an environment with so many non-judgmental friendly people." We have come to see these "soft" outcomes as central to our purposes.

Seek Out Wild Locations Near and Far

A single question will dominate early conversations about getting students out of the classroom: Where to go? The options will obviously vary widely depending on geographical location. Dawson's Montreal campus boasts a significant green space with mature trees and gardens and is located a 15-minute walk from several large downtown parks. Other significant urban greenspaces are accessible by public transport, and parks and wildlife refuges are nearby just off-island. Regions to the immediate north and the south of the city offer more remote locations for single and multi-day hiking, camping, and canoeing. These options have served different purposes. We use the campus greenspace, for example, for shorter activities that are

FIGURE 8.1 Students taking a break for journaling at lac du Poisson-Blanc.
Source: Ian MacKenzie.

integrated into class time. When Doug introduces camping equipment and set-up (tents, camp stoves), he distributes and explains the gear in class, and then we shift to the campus greenspaces for demonstrations and practice. When Ian investigates the history of field guides, he introduces Audubon, Peterson, iNaturalist, and Merlin in class, and students then move outside to practice identifying plants, trees, and birds local to the Dawson campus. Nearby parks are destinations for longer two to four-hour excursions. Taking advantage of the back-to-back scheduling of the classes, we have sufficient time for multiple learning activities. Because these greenspaces are embedded in the urban built environment, they also provide opportunities to point out the tension between the human imperative to control and dominate and the self-willed character of natural ecosystems.

Alongside out-of-class excursions on campus and to local parks, the outdoor education intensive includes a total of 18 hours (or three full days) for off-campus trips. Mandatory components of the Physical Education coursework, we plan these trips well in advance of fall registration and spread them across one, two, or three weekends, depending on our objectives. For example, a single three-day trip permits the use of more remote wilderness locations for camping and canoeing. Once we confirm locations, we factor them into decisions on topics, readings, and learning activities.

In fall 2020, because of COVID restrictions and the move to online classes, we allocated the intensive trip hours to three solo trips, planned autonomously by each student under Doug's guidance and executed on designated weekends at locations in the Montreal region. This change introduced an entirely new set of challenges, to which Doug responded with a detailed process of comprehensive trip proposal and approval; day-of-trip check-ins and check-outs via text messaging; and trip journal submissions afterwards. The result was certainly different from our intense group trip experiences, but on the other hand, after each trip day, our Zoom classes were animated by a rich variety of experiences and stories, as students reported on their day's activities and journaling insights. Finally, in 2021, with a return to in-person learning but travel to remote locations prohibited by college COVID policy, we planned three separate days of activity over the term, using several on-island Montreal locations.

It is hard to exaggerate the impacts of these trips on students. The successful recipe, we think, boils down to three simple steps: Strenuous and/or intellectually challenging outdoor tasks; reflective journaling on related texts and thematic ideas; and sharing in group discussion. These steps permit students to articulate and value their new competence in these tasks and affirm their achievements with each other. When the trip ends, students can confirm in their own minds, "I did it," and also, "We did it." One student reflected, "It was great! We laughed together and made new memories together. I didn't think that camping with my teachers would be fun, but I was wrong! I never actually liked camping before, but they showed and taught us new things about how to survive outdoors, which ended up by being an unforgettable experience for me."

Create a Framework for Integration

Our first iteration of "Nature Revisited" was somewhat incoherent on the actual integration of English and Physical Education curriculum. While specific learning activities and trips were successful, we lacked a clear, over-arching learning narrative uniting the two courses. We needed an explicit conceptual framework that would facilitate a tighter integration of the two courses. Subsequent reading led us to a variety of resources for formal and informal outdoor learning. We drew on David Sobel's seven design principles for educators (Sobel 2008); the six "touchstones for wild pedagogues" of Jickling et al. (2018); and especially on Jon Young's thirteen routines for nature connection (Young et al. 2016). We adapted some aspects of each to the creation of our own framework, mapping twelve outdoor "core practices" onto a set of complementary topics and themes in English.[1] As a result, students now read about a skill or mindset in English, and in short order practice it in the Physical Education class and on the intensive trips.

To understand how a core practice might structure learning activities in both courses, let's look at wilderness readiness as an example. In the second week of the semester, students read the classic Jack London story "To Build a Fire" for Ian's English course. They investigate the setting of the Klondike gold rush, and analyze features of the adventure genre. The protagonist is walking a solitary winter journey between frontier camps on the Yukon River, accompanied only by a dog and

TABLE 8.1 Core practices shared across outdoor education and English curricula.

Core practices	*PE outdoor education topics*	*English topics*
Outdoorsy—or not?	Assessing prior nature experiences and attitudes; identity, culture, nature; fitness self-assessment; goal-setting	Storytelling, oral and written traditions; definitions of nature, wild and wilderness; diversity and inclusion
Connecting to place	Mindfulness and placefulness; landscape and ecosystem awareness; cardinal points; leave no trace	Contemplative traditions in nature writing; the sublime in art and literature; the sense of wonder
Walking and exploring	Walking and hiking techniques; human evolutionary history and bipedal biomechanics	Walking and travel narratives; walking and poetic meter; nomadism, song and story
Wilderness readiness	Basic needs (shelter, water, food); risk management; trip planning and decision-making; gear, tools and tool making	Adventure plots and genres; character and nature settings
Sensory awareness	Human physiology; enhancing sensory awareness	Sensory description and diction; nature imagery and symbolism
Nature-inspired play and creativity	Outdoor play and games	Nature art sculpture, soundscapes
Curiosity	Nature-based questions and interpretation; landscape features, tracking, scat, markings, plant and bird identification	Field guides and natural history; field apps; conservation and citizen science
Nature journaling	Reflective journaling and meta-cognition; mindfulness and health; sketching	Observation and journals in the natural sciences; contemporary nature journaling
Animal movement and communication	Animal physiology and movement: fox walk, owl ears, hawk eyes, wolf smell, fight/flight movement; bird songs and calls	Animals in poetry and story; interspecies communication; human-animal transformations
Compass and mapwork	Orienteering, map-reading and compass skills	Landscape and memory; childhood secret places; neighborhood and ecosystem map-making; literary maps real and fantastical
Individual and social well being	Nature and mental health; responses to eco-anxiety; nature deficit and nature access; environmental health	Poetry and narratives of trauma and recovery; diversity in outdoor activity; environmental justice
Gratitude and agency	Gratitude mindset; environmental ethics; stewardship behavior	Eco-spirituality and wisdom; scientific and traditional Indigenous ecological knowledge; poems and narratives of gratitude and benediction; writing for enviro-activism

facing –50°F temperatures (roughly –45°C). After breaking through thin river ice and soaking his legs, he needs to quickly build a fire to stave off hypothermia. The story vividly dramatizes what it means to be alone and exposed to the cold and how essential a fire at that time might be. Events at the crux of the story show how a seemingly straightforward process of fire-making can go terribly wrong, and student discussions focus on human arrogance, vulnerability to harsh landscapes, and chance.

A week later, during the first weekend intensive trip, we task students with quickly building and starting a small one-match fire. In the set-up for the activity, Doug surveys the basics of fire-making with students, including safety, the suitability of materials at hand, methods of construction, and leave-no-trace clean-up. Ian also prompts students to recall specific aspects of the London text: "Like him, you are cold, you are suffering from hypothermia, you are alone, and you absolutely need a fire for your survival. What was he thinking and how did he react? What are you thinking and how will you react?" On the second day of the 2017 trip, these questions were not difficult for students to assimilate, as on the first night, temperatures had fallen below 0°C and it had snowed! Students proceeded in groups of three to build and start their small fires, laughing nervously about freezing to death if they failed. All survived—but the experience remained a topic of discussion throughout the rest of the trip, with students referring to both their fires and the story. This is the kind of integration that we were hoping for. Helping students link the story with their own fire-starting efforts informed their reflections on their own skills and vulnerabilities, and on their presumptions about human dominance over the powers of the natural world. This framework has been perhaps our single most important curricular innovation, permitting us to realize an integrative sequence of topics and learning activities for both courses for the entire semester. Student feedback reaffirmed the value of this structure:

> I was blown away with the results of these paired courses. They complimented each other beautifully, learning in one course helped us learn in the other and also the reverse. The theme of nature was extremely intriguing, learning about nature, different animals, sounds, and how we relate to nature itself was truly interesting. The cherry on top was actually applying this knowledge in the real world. It deepened the connection between myself and nature, not only did I learn about nature but I *experienced* it.

Take Play and Creativity Seriously

In *Wild Pedagogies*, Jickling et al. (2018) argue for outdoor learning experiences that escape the control paradigm of institutionalized education. Standardized exams and testing, prescribed learning outcomes, ranking of students via R-scores or GPAs—these features of an industrialized model of learning, in their view, indicate how deeply we have allowed the flourishing of human natures to be colonized by a reductive set of means and ends. It may not be possible to escape entirely the control imperative, they acknowledge, but it is within the power of outdoor educators to resist it by "embracing complexity, inviting risk and allowing for emergence" (ibid., p. 86).

FIGURE 8.2 Students create a nature art project on Île Saint-Bernard.
Source: Ian MacKenzie.

We respond to this injunction by first making a connection to the idea of play and its central role in learning in complex organisms. Many of our core practices overlap with familiar forms of playful outdoor behavior. According to Gray (2013), play is activity that is self-chosen and self-directed; intrinsically motivated; guided by mental rules; imaginative; and conducted in an active, alert, but relatively non-stressed frame of mind. For Doug, this definition permits a critique of the prescriptive, sometimes controlling nature of his default pedagogy. In his view, the discipline of Physical Education is sometimes beset by an over-eagerness to be taken seriously, perhaps because of a lingering insecurity that the discipline borders too closely on the forbidden kingdom of playful fun. Doug felt that he was often assigning procedural tasks whose outcomes were pre-determined and, thus, easily quantifiable, when these tasks were contrary to the free, intrinsically motivated development of skills and aptitudes that he wanted. For example, teaching both wandering and animal movement would ideally create moments for students to explore spontaneously, to react unscripted to their senses in real time, and to draw their own conclusions about the experience without any fear of there being a wrong answer. Upon arriving at a trail intersection, students might be asked to suspend any thoughts about the right direction, and instead let curiosity and instinct choose their next direction of travel. As one student commented:

> Wandering was a bit challenging because I am a person that likes to plan because it makes me feel in control. However, it felt good just going where your body tells you to, and during my intensive days, I found myself doing only that. I would just follow a sound or a bird or my feelings and I felt free.

For Doug, this kind of feedback has confirmed the value of letting go of his disciplinary doubts and trusting students to apply the core practices in their own ways, following their curiosity and intuition.

We inject more creative spontaneity in the form of a Nature Art project, a multi-stage learning activity and assessment in the English course. Ian familiarizes students with two artists whose processes model a responsiveness to the natural world: Scottish landscape sculptor Andy Goldsworthy and American soundscape composer Cheryl Leonard. These artists practice sensory awareness in specific natural settings (e.g. for Goldsworthy, the Scottish moors; for Leonard, melting Antarctic glaciers) and then channel that sensory input into the creation of sculptures and sound compositions of extraordinary beauty, using found materials and remixed field recordings. Students are challenged at first by the idea that, like Goldsworthy, they will fabricate a sculptural installation in dialogue with a specific locale with nothing but found materials. In a multi-stage process, they select a promising location; note its physical geography and available materials; write a detailed proposal; and then in a few hours on a specified day, gather their materials, prepare their site, and make art. They document their work with images and an artist's statement, which are published to the course blog. Students finish by loving this project, which gives them an opportunity to create something personal and original in dialogue with their chosen physical landscape.

Scaffold Deep Learning through Writing

We are convinced that learning new outdoor skills and mindsets can be enhanced by careful design of scaffolded writing activities, not only in English but also in Physical Education. We see this confirmed in the quality and frequency of student contributions to Ian's course blog and in higher-order reflection in Doug's journaling assignments.

The bulk of student course work for "Into the Wild" takes place on a course blog, a virtual environment which is ideally suited to scaffolding writing and giving effective feedback. Students publish to the course blog weekly informal reading responses, revised formal assignments, and optional special interest "Wild Things" posts. The blog is central to the "community" aspect of our learning community, growing over the term into a rich ecosystem of thinking and writing. When assignment "submission" is largely replaced by publication, a class becomes an authentic community of writers. Blogging exchanges the private two-way street of student-teacher submission and evaluation (Giltrow 2012) for an animated neighborhood of conversations. Informal response writing feeds into revised formal products. Ian's surveys confirm that confidence in writing grows rapidly in the blog's public light. Students develop a collective ethic with respect to the publications of the community and student success is fostered by peer feedback and models. Blogging also has an important aesthetic dimension. Posts can be creatively designed with the addition of nature-related images, video, and sound, and digital visual literacy can play a supporting role in motivating high quality student thinking and writing.

On the intensive trips, students complete journal entries for Physical Education where they reflect on their experiences with the core practices. In the fall 2020 and 2021 COVID semesters, students did three separate days of activity and journaling, which allowed Doug to scaffold his prompts and feedback. For example, in preparation for the first trip, Doug introduced Bloom's taxonomy and the distinction between different levels of cognitive activity. He felt that this meta-cognitive framework might help students to evaluate their journal entries about their quiet connection activity. Subsequently, when offering feedback on the first submission of journals, Doug indicated that many students described the physical aspects of their sit spot location and their behavior and stopped there. Following Bloom's scale, this first effort suggested that students remained at the level of Understanding, in that they described, discussed, and identified their experiences.

For their second and third journal entries, Doug encouraged students to develop their thinking at the higher levels of Analysis and Synthesis. Referencing several of the English readings for this core practice, he asked students to delve more deeply into both external and internal phenomena and to hypothesize relationships and meanings that connected those phenomena. One student wrote:

> The quiet connection place was a big part of this change/evolution in my health because it felt like a safe place. I liked that I could go there alone and that not a lot of people knew about the location. I could really escape all my worries about school that made me feel trapped in the house. I used to hate being in my room because I associated it with school so much that it gave me anxiety. Whenever I felt this way, I would imagine my sit spot, I would imagine the wind on my face, the sound of the waves hitting the shore of the little island, the sounds of the birds around me and me sitting on that rock. It really brought me peace because this place is associated with relaxation, mindfulness, and meditation. The sit spot also made me feel more connected with nature, as I said in one of my past missions, I felt as if the little island and I made one. I felt like we were connected, I felt like the birds, squirrels and other animals did not mind my presence and it felt great.

The positive role that this reflection and realization might have played in helping the student traverse month after month of isolation and stress during the COVID pandemic seems self-evident.

Cultivate Diversity and Inclusion in Outdoor Learning

How does learning about and practicing connectedness to the natural world intersect with advancing social justice? The events of the past two years have forced us to address this question head on. Montreal is a diverse city, multicultural and multilingual, and Dawson is an English language institution in a French language majority province. Our student population embodies a complex social reality, and as our English and Physical Education classes are general education requirements, they represent a true cross-section of that reality. Leveraging the diversity of our students has long been an important part of building a positive learning environment. While some students come to us with extensive outdoor experience, many do not. Whether due to economic or socio-cultural factors or simply preference, nature-based recreation can involve unfamiliar and sometimes uncomfortable experiences. Thus, the first core practice, "Outdoorsy—or not?" is devoted to examining students' prior experiences and validating the full range of responses, both positive and negative. We encourage students with little outdoor experience to listen to those with more and consider how they have successfully entered unfamiliar environments and learned new skills in the past, often with the assistance of a community of supportive peers. In a practical sense, students must overcome their reservations about working with others. On the trips, camaraderie gels in group activities and debriefs and in conversation and games at night around a fire, when we share stories of success and failure with liberal outbreaks of laughter. A key insight for students is the moment when they realize that connecting to nature also means connecting with and appreciating each other—over and above their differences.

For the fall 2020 version of "Nature Revisited," we knew we had to do more. We could not fail to speak to the Black Lives Matter movement, for example. On May 25, 2020, Minneapolis police officers murdered George Floyd. On the same day, a white woman falsely accused New York birdwatcher Christian Cooper of assaulting her in Central Park. As we prepared for the fall, Ian was determined to start a frank exploration of diversity and racial exclusion in outdoor culture. Thus, in the first full week of the semester, students read and wrote about an interview with Dr. Carolyn Finney ("Landscapes of Exclusion") and a *New York Times* feature interview by Sarah Nir with Christian Cooper ("The Bird Watcher, That Incident"). We carried the topic forward as a series of open questions for discussion through the rest of the semester: How are concepts of nature and environment inflected by cultural assumptions and prejudices? How can outdoor activities and spaces be made more inclusive and welcoming? How can a love of the outdoors open a space for conversations and actions about racial justice and reconciliation?

Equally, we were compelled to address the topics of Indigenous ecological knowledge and the colonizing premises and viewpoints embedded in debates and actions on wilderness and conservation. This is a necessary and urgent topic, one that faculty must broach with students and confront among themselves. In Canada, the role of Indian residential schools in a larger national project of genocide was thrust into the headlines in 2021 by the discovery of thousands of unmarked graves at residential school sites across the country—the unmarked graves of children who were taken from their families, their communities, and their traditional lands and cultures. In response, we launched a process of decolonizing our own ideas and attitudes about "our" outdoors. There was a single token representation of Indigenous writers in Ian's early syllabi, as one student pointedly remarked in her end-of-semester feedback on the course. We took this as a literal indictment. We were a pair of white settlers, rhapsodizing about the great outdoors over an erasure of the realities of how those spaces were first "settled" by white Europeans; exploited for their exclusive economic gain; and later cordoned off and "conserved" for their recreational pleasure. The creation of North American "wilderness," as William Cronon observes in "The Problem with Wilderness," was largely premised on the displacement of Indigenous peoples from their traditional lands and the destruction of their cultural ties to those lands. In fall 2020, we added Winona LaDuke's seminal "Traditional Ecological Knowledge and Environmental Futures" to the English syllabus, along with several chapters from Robin Wall Kimmerer's *Braiding Sweetgrass*. It shouldn't have surprised us that some of our students were much further down this path than we were, and we ended up drawing on their knowledge and following their lead in discussions. We are at the beginning of a long process, and have much to learn.

Conclusion

The six principles explored here have guided us in the creation of opportunities for students to directly experience the beauty and complexity of wild places; to think and write about the natural world and their relation to it; and to share their experiences, both intellectual and embodied, in a supportive community that values their learning. The student community supports not just intellectual growth but emotional depth in the connection to more-than-human nature. While the "sense of wonder" might be sparked as a solitary intuition, it becomes infectious as it is shared and affirmed. Last, but certainly not least, our own experiences suggest that outdoor learning communities are professionally rewarding for teachers. Because of the atmosphere of familiarity and trust, we have grown comfortable sharing pedagogical insights with each other, and sharing our deep love for wild places and wild, non-human kinfolk with our students. As equal members of our various "Nature Revisited" cohorts, we have only deepened our love of the outdoors, and of learning together outdoors.

Acknowledgment

Dawson College is located on land which is the unceded traditional territory of the Kanien'kehá:ka. This land has also served as a gathering place for Abenaki, Anishinaabe, and other nations. Indigenous peoples are the original title-holders of these lands and waters. It is not enough to only acknowledge the keepers of the land. It is our responsibility to become informed and to get involved in anti-colonial and

Indigenous struggles and resistance. Wherever Indigenous peoples are protecting and reasserting control over their territories today, they are at the frontline of our collective struggle to protect and steward our planet.

Note

1 For a list of our core practices and complementary topics and themes in English, see the companion website.

References

Cronon, W., 1996. The trouble with wilderness: Or, getting back to the wrong nature. *Environmental History*, 1 (1), 7–28.

Giltrow, J., 2012. *Landscaping with indigenous species: How to replace quizzes, exams, term papers (non-native species) with species native to the research culture.* Keynote address, McGill University Writing Conference.

Gray, P., 2013. *Free to learn: Why unleashing the instinct to play will make our children happier, self-reliant, and better students for life.* New York: Basic Books.

Jickling, B., *et al.*, 2018. *Wild pedagogies: Touchstones for re-negotiating education and the environment in the Anthropocene.* Cham, Switzerland: Springer.

Kimmerer, R. W., 2013. *Braiding sweetgrass.* Minneapolis, MN: Milkweed Editions.

Kuh, G. D., 2008. *High-impact educational practices: What they are, who has access to them, and why they matter.* Washington, DC: Association of American Colleges and Universities.

Laduke, W., 1994. Traditional ecological knowledge and environmental futures. *Colorado Journal of International Environmental Law and Policy*, 5, 127–148.

Nir, S. 2020. The Bird Watcher, That Incident and His Feelings on the Woman's Fate. *New York Times*, May 27, 2020.

Smith, B. L., *et al.*, 2004. *Learning communities: Reforming undergraduate education.* Hoboken, NJ: Wiley.

Sobel, D., 2008. *Childhood and nature: Design principles for educators.* Portsmouth, NH: Stenhouse Publishers.

Wabuke, H. 2015. Landscapes of exclusion: An interview with Carolyn Finney. *Guernica*, September 15, 2015.

Young, J., Haas, E., and McGowan, E., 2016. *Coyote's guide to connecting with nature.* Shelton, WA: Owlink Media.

Zhao, C. M., and Kuh, G. D., 2004. Adding value: Learning communities and student engagement. *Research in Higher Education*, 45, 115–138.

9

FROM DINOSAUR TO BEARS EARS

Engaging Utah's Public Lands via Interdisciplinary Inquiry and Experiential Learning

Judson Byrd Finley and Keri Holt

Few issues engender more consternation among citizens of the rural western United States than who maintains the rights of ownership, access, and use of America's federal public lands. Most public lands are in the West where tightly knit rural communities share generations of cultural and economic history derived from the labor of ranching and mining founded on an even earlier legacy of exploration, colonization, and dispossession. Many rural citizens remain deeply suspicious of outsiders who may threaten the continuity of their cultural traditions and economic prosperity derived from the land itself. Nothing draws that ire more than the perceived overreach of a distant and disconnected federal government and an overprivileged and even elitist conservationist movement. Should national conservation interests prevail over local economic gain? Or do we cede territory in the interest of American energy independence and domination? Whose voices are heard in these debates? Whose voices are not?

These conflicts and questions have become particularly visible in debates surrounding Bears Ears National Monument. In 2016, under the authority of the Antiquities Act, President Obama established Bears Ears National Monument to "preserve its cultural, prehistoric, and historic legacy and maintain its diverse array of natural and scientific resources" (White House 2016). Although Obama's proclamation had strong support from Indigenous groups, conservationists, recreation enthusiasts, and scientists, those who opposed it saw the monument as a dangerous act of federal overreach that hampered economic development. This controversy intensified in 2017 when President Trump revised the monument boundaries and reduced its size by 85 percent, resulting in widespread demonstrations and media campaigns voicing reasons to support, revise, or reject the monument.[1] With so many different opinions, perspectives, histories, and proposals involved, the Bears Ears debate is unbelievably complicated and divisive. But as Friends of Cedar Mesa director Josh Ewing notes, "People working together bridge gaps. We may never

DOI: 10.4324/9781003221807-9

agree on certain things, but we can agree to work on others" (Robinson 2018, p. 148). Citizens need to learn how to understand and engage in these debates, particularly young people who will be living, working, and contending with the future impacts of public land management.

We describe an interdisciplinary and field-based curriculum that is designed with strategic intention for empowering college students to recognize themselves as stakeholders in public lands conservation. We teach at a public land grant institution in northern Utah where public lands debates are continually in the news, not only in reference to Bears Ears but also to matters related to road use, fossil fuels, water rights, and environmental policies. Despite the visibility and importance of these debates, many of our students know very little about the issues, and those who have some knowledge either don't know how to participate or feel that their voices don't matter to the final outcomes. How do we engage our students in a way that simultaneously informs them about these debates while empowering them as stakeholders in a complex process?

Our class is designed to create a cadre of citizen-scholars who can confidently position themselves in public lands debates. We specifically focus on the public participation process federally mandated by the National Environmental Policy Act (NEPA), teaching our students about its intentions, structure, and complexities with the goal of enabling them to participate competently within it. To illustrate the importance of NEPA and the impact of public engagement, we organize the course around two different case studies—Dinosaur National Monument and Bears Ears National Monument. Understanding the varied uses and conflicts over land management within these monuments requires interdisciplinary perspectives, and we draw on the methods of Anthropology, History, Literary Studies, Archeology, and Environmental Science, combined with critical analyses of the socioeconomic and physiological impacts of natural resource development, to help students understand the seemingly abstract "human environment" that NEPA regulates. While this interdisciplinary approach helps our students understand the multiple uses and stakeholders that NEPA addresses, fieldwork and experiential learning are essential tactics for connecting students to these issues and motivating them to participate in public land debates. These joint features—interdisciplinary study and experiential learning—are crucial methods for helping students grasp and negotiate the complexities that inevitably emerge in all public land debates, while also empowering them to take part. In our experience, this method produces the active and engaged advocates we think are necessary to the future of American conservation.

NEPA as a Pedagogical Polestar

Since teaching students how to engage with NEPA's public participation process is a core course objective, we begin with a brief introduction to this law and its procedures. Most students are unfamiliar with NEPA, so we open with a lecture outlining the history and key features of this federal act. NEPA is best understood as a "Stop, Look, and Listen" law (Advisory Council on Historic Preservation

2013) that guides decisions about the environmental impacts of development projects involving federal funds, lands, or permits. The law was a reaction to the environmental consequences of unbridled American economic growth during the twentieth century, epitomized by the burning of the Cuyahoga River. The core tenet of NEPA is for the federal government to take into account the effect of its actions on the quality of the human environment. The "human environment," however, encompasses a nexus of natural, social, and economic factors that occupies a contested space between individual economic gain and collective well-being. In practice, NEPA is the "umbrella" under which falls a network of federal laws for managing natural and cultural resources, including the Clean Air Act (1970), the Clean Water Act (1972), the Endangered Species Act (1973), and the National Historic Preservation Act (1966), among other "common sense" policies (King 2013). Thus, NEPA is interdisciplinary by design, and engaging with the NEPA review process helps our students recognize the value and necessity of examining the use and management of public lands from the range of perspectives we explore in the course.

In addition to its interdisciplinary scope, NEPA is a guiding pedagogical tool for our course because it creates a space for public dialogue. We work with our government documents librarian to introduce students to the Federal Register, teaching them how to identify and track NEPA review for ongoing projects. NEPA requires an extensive review process to identify the scope of potential impacts, provide alternatives for project implementation, and propose mitigation strategies for the preferred alternative. Two simple guiding tenets in this NEPA review are, "What is at stake?" and "Who are the stakeholders?" In other words, the job of the federal government in the NEPA review process is to stop (i.e., take into account the effect of its actions), look (i.e., identify what is at stake), and listen (i.e., recognize and acknowledge stakeholder concerns). We remind students that nowhere in either the title or guiding language of NEPA is *protection* made explicit. Instead, "policy" implies procedure, procedure creates a place for participation, and participation means advocacy. Of course, the NEPA review process is time consuming and often inaccessible. Despite these challenges, we remind students that one of its key outcomes is to create a process for civic engagement.

As we turn our attention to the requirements and intricacies of the NEPA review process, one of our primary learning objectives is to inspire students to be effective NEPA conversationalists. The prospect of taking part in these public reviews excites our students, but their lack of knowledge about the challenges of ensuring equitable and accessible participation and the procedural difficulties of the process itself frustrates them. To address this, we task them with developing a handbook to explain the NEPA process and teach the public how to negotiate its complexities in user-friendly terms.[2] Arriving at this final outcome requires first teaching them about the varied uses, stakeholders, and impacts involved in public lands debates and the interdisciplinary perspectives needed to approach them, which brings us to our case studies of Dinosaur and Bears Ears National Monuments and the field trip that ties the components of our course together.

From Dinosaur to Bears Ears: Situating the Public Lands Debate

To provide context for understanding public land controversies and the value and impact of public engagement, our course juxtaposes the history of land use, management, and conservation in Dinosaur National Monument with the ongoing debates in Bear Ears. Dinosaur was established by President Wilson in 1915 following the lobbying efforts of paleontologist Earl Douglass, who discovered a major fossil bonebed in the remote desert near Vernal, Utah. Drawing on the relatively new Antiquities Act, Wilson established an 80-acre monument preserving the site, where dinosaur fossils remain a central attraction. President Roosevelt expanded the monument in 1938 to encompass the wild and scenic river corridors of the Green and Yampa Rivers and protect the rich archaeological history of Indigenous agricultural societies visible in its expansive rock art galleries. The region remains central to the traditional homelands of the Uintah and Ouray Tribe of Ute Indians, and the monument also preserves a varied history of Euroamerican exploration and settlement, beginning with the 1776 Domínguez and Escalante expedition and continuing with the early American fur trade and the scientific explorations of John Wesley Powell who successfully navigated the Green and Colorado Rivers in 1868. During the late nineteenth century, the remote and challenging features of the landscape attracted settlers who bucked convention, exemplified by the colorful legacy of the Bassett sisters, who ran successful ranching operations in remote Browns Park, and outlaws such as Butch Cassidy's "Wild Bunch," who used the region's rugged terrain to avoid the law. This adventurous landscape eventually became popular for recreation in the twentieth century, activities that have had to coexist with the widespread development of mining and oil and gas industries. We study this long and varied history using exploration journals, newspapers, federal reports, archeological and scientific studies, interviews, photographs, and even film clips. These primary sources help illustrate the many ways that people have used these lands and how this history influences their use and management today.

At the center of our Dinosaur study is the 1950s controversy over the Echo Park dam, which sparked the modern environmental movement (Harvey 1994). In the 1950s, the Bureau of Reclamation proposed dam sites within Dinosaur along the Green and Yampa Rivers to regulate water supply and increase the region's economic potential. The proposed dams, however, would effectively drown the Dinosaur river corridors and their rich natural and cultural heritage, prompting significant environmental, cultural, and recreational consequences while further disaffecting Indigenous communities. These outcomes were particularly controversial because they were located within the boundaries of a national monument. Since the Echo Park dam was proposed before NEPA, there was no formal opportunity for the public to participate in the proposal process. The Sierra Club launched an aggressive public relations campaign, exemplified by the publication of *This Is Dinosaur,* a collection of essays edited by Wallace Stegner (1955) that was sent to all members of Congress. Drawing on the perspectives of archeologists, geologists, hikers, river runners, historians, and journalists, this book became an important catalyst for prompting public engagement with federal lands policy by highlighting the varied uses of the monument and how the dam would negatively impact them. In addition to studying this book's impact, we also take students into our university library archives to explore arguments that supported the dam, examining how people used local and national media to promote the dam through posters, rallies, pamphlets, and editorials (Jensen 2018). By introducing students to these varied stakeholders and the impact of their public engagement, the Echo Park dam controversy lays the groundwork for examining the highly regulated yet equally controversial decisions surrounding the management of Bears Ears, which is the focus for the second half of our course.

The land encompassing Bears Ears has an equally long and dynamic history, which we approach by first identifying key similarities between the two monuments. Like Dinosaur, Bears Ears has a rich Indigenous history. The region is home to thousands of ancestral Puebloan, Ute, and Navajo sites, and the area continues to be an important spiritual site for members of the Navajo, Ute, Hopi, and Zuni Nations. Spanish and Anglo-American exploration and settlement, especially regarding the cultural and economic influence of Mormon settlers, has also shaped Bears Ears. Like Dinosaur, Bears Ears has primarily been used for ranching, farming, and fossil fuels extraction, with a more recent recreation and tourism boom. These industries exist in tension with efforts to protect the region's natural landscapes and cultural and historical legacy. Identifying the similarities helps our students reflect more deeply on the challenges of balancing multiple uses and the interests of varied stakeholders, challenges which dominate the Bears Ears conversation.

Our shift to Bears Ears introduces several issues not in play at Dinosaur, heightening the importance and challenges of fostering effective public engagement. First is the intense politicization surrounding the creation of Bears Ears. The president's power to establish national monuments through the Antiquities Act has always been controversial, but the divisive response to Bears Ears indicates just how contentious these designations have become. President Trump's decision to reduce the monument's size intensified these arguments, making it a highly publicized site for debating the limits of presidential power and determining the role of state, federal, and Indigenous authorities in regulating public lands and balancing conservation and economic interests. Positions on these issues tend to follow partisan lines, which means that discussions of public involvement require us to negotiate political affinities that can be challenging to address in the classroom.

Although we want our students to be aware of these political differences, we don't want to foster a hostile or divisive classroom environment. We continually remind our students that our class is a space to listen and learn about the different stakeholder views, not to resolve them. Learning about the varied and complex perspectives, rather than finding solutions, helps students approach these discussions with an open and noncombative attitude. We also carefully select a textbook that models this approach. The collection *Voices from Bears Ears: Seeking Common Ground on Sacred Land* (Robinson 2018) comprises interviews with diverse stakeholders in the Bears Ears debate highlighting an array of "competing visions," while simultaneously foregrounding the "individual voices and personal histories" that inform those views (Robinson 2018, pp. xviii–xix). This book invites readers to reflect on the varied positions, rather than judge or dismiss them, a framework that allows us to avoid evoking partisan tensions and animosity in the classroom and, instead, foster an awareness of the complexities that lead to these conflicts in the first place.

Another key difference when comparing the history of Dinosaur and Bears Ears concerns the role of Indigenous stakeholders. The establishment of Dinosaur and the Echo Park dam debates did not include the local Ute tribe. In contrast, the Bears Ears process has been deeply engaged with Indigenous stakeholders. The Bears Ears Inter-Tribal Coalition, a collaborative Indigenous advocacy group composed of Hopi, Navajo, Ute Mountain, Zuni Pueblo, and Ute tribal members, supported its establishment, and it is co-managed by representatives from these five nations along with the Bureau of Land Management and Forest Service. Not all tribal members support the monument, however, and strong disagreements remain within Indigenous communities about the status and use of the monument, as well as ongoing tensions between Indigenous, federal, and state officials over the terms of their collaborative management. Examining the varied Indigenous stakeholder perspectives gives our students a more inclusive sense of this issue, which, in turn, magnifies their awareness of the emergent divisions and complexities.

The Indigenous experiences and cultural history are often unfamiliar to our students, requiring additional instruction to help them approach these perspectives in informed and responsible terms. We approach this through short lectures that provide background about Indigenous histories, as well as readings about US federal Indian policies and their impact on Indigenous communities, particularly regarding health care, education, and employment. By far, however, the most effective means of introducing students to Indigenous history and culture are the interviews and personal stories of Ute, Ute Mountain, Navajo, Zuni, and Hopi people featured in *Voices from Bears Ears: Seeking Common Ground on Sacred Land* (Robinson 2018). These experiences, delivered in a personal, first-person narrative, provide our students with a powerful introduction to the Indigenous perspectives at stake in this debate.

Finally, Bears Ears also differs from Dinosaur regarding the public's role in evaluating and responding to the federal policies for managing it. While the Echo Park dam controversy enabled our students to see the importance of including the public in decisions about how to use and manage public lands, the history of that movement also illuminates the limits of relegating involvement solely to local and national media. In the Bears Ears case, proposals regarding the use and management of lands within the monument are subject to NEPA review, giving the public a direct and active role in the development of federal policies. Teaching our students about the objectives of NEPA and how to participate in its public processes is of central importance in our study of Bears Ears, particularly since many of the polices for managing this monument are currently in development or up for debate.

These three factors—the divisive politics of Bears Ears, the engagement of Indigenous stakeholders, and the complexities of the NEPA process—are challenging topics we address in the second half of the semester. While our study of Dinosaur National Monument lays a vital foundation for understanding the conflicts and tensions surrounding Bears Ears, we need to find a way to motivate our students to engage with these complexities and see themselves as stakeholders in the debates. To do so, we turn to experiential education, bridging the two halves of our course with a three-day field trip to Dinosaur National Monument. By experiencing this landscape directly and responding to its current uses in conjunction with their knowledge of its history, our students no longer see themselves as distant observers but as engaged citizens who are not only affected by the way these lands were used and managed, but who also feel they have a responsibility to take part in determining their future.

Reading Dinosaur: A Field Guide for Interdisciplinary Studies

The genesis of our class was a simple conversation between a literary historian and an archaeologist about our shared interest in Dinosaur National Monument, the many ways that we can "read" its landscape, and how important it is for students to see *and* feel the places we teach about. While the interdisciplinary pedagogical approach we describe creates the requisite framework for intellectual diversity and a context for civic engagement, we find that field experience is crucial for building community, investing people in places, and empowering students as stakeholders in their use and management. Our primary objective is to transform classroom learning into real-world experience. We must bring to life the historical and make real the natural while challenging students to strike a balance between conservation and economy—the core principle of NEPA. Our three-day field trip to Dinosaur is designed with the strategic intention of transforming abstract classroom learning into experiences that empower students as citizen scholars and stewards of our nation's natural and cultural legacy.

Our field experience is a vehicle-based camp trip that occurs at the midpoint of the semester. The itinerary and lesson plan focus on the history, science, and conservation of Dinosaur that seeks balance with a local economic legacy of ranching, mining, and fossil fuel extraction.[3] We adapt the concept of "close looking" from Art History (Specter 2017) that encourages students to make careful observations of the landscape and at our stops along the way. We provided students with a field journal, where, in addition to a space to complete all the field work assignments, we encouraged keeping track of any observations they made throughout the trip.

The first leg of the trip is through rural southwestern Wyoming, a vast open space of sagebrush desert with many active and inactive gas wells, a few antelope, and fewer people. This part of the trip challenges students to consider how the land supports people today and what might matter to their economy. Our route passes the place where American explorers described the Green River entering the abyss of Red Canyon with its whitewater now submerged beneath Flaming Gorge

Reservoir. During our first stop at the Red Canyon overlook, students consider how the still waters of the reservoir compare with the writings we studied by Frémont, Powell, and other early explorers who described a wild river that no longer exists due to the legacy of water storage and hydroelectric power generation. As we pass through the city of Vernal, Utah, a community defined by the boom-and-bust cycles of fossil fuel extraction, we ask students once again to consider what drives the local economy and what their concerns might be as stakeholders in land-use debates. This is an important point of reflection for voices in opposition to conservation in national monuments like Dinosaur and Bears Ears. We reach Dinosaur in the afternoon of the first day. The geology of Split Mountain and the local landscape of the Green River is profound as it exits the Uinta Mountains. We travel into Cub Creek, the site of an Indigenous farming community active 1,500 years ago, and our second major stop is a rock art gallery from this ancient time. The complex imagery of the rock art challenges students to think carefully about the lives of the people that lived in this high desert and to hear the echoes of an otherwise silent archaeological record. Not far from this gallery is our next stop, where Josie Morris, a first daughter of American settlers to Browns Park, continued to live in a dirt-floored log cabin until her passing in 1964. Josie's Cabin juxtaposes Indigenous and historic lifeways while provoking students to consider life in a time not so distant from their own. Students camp on the banks of the Green River in the very place that John Wesley Powell's crew took their respite.

Day two begins with a focus on geology as students are asked to "read" the complex geological record. What forces brought changes to this landscape in distant eons? We pose a similar question at the Carnegie Quarry Hall, the attraction of Earl Morris Douglass's envisioned monument. Again, we ask students to follow principles of "close looking" to "read" the death site of dinosaurs 150 million years ago. From the Carnegie Quarry we travel to the Canyon Country, which encompasses the confluence of the Green and Yampa Rivers and its centerpiece, Echo Park. The hike to Harper's Corner overlook is little more than a mile, but here students stand on the very spot where David Brower and Wallace Stegner wanted them to witness canyon country unblemished by a steel-reinforced concrete dam and the flat, impounded waters of two once-wild rivers. Harper's Corner is the most-wild classroom for us. It is a place to remind our students that stakeholder opinions for and against developments like the proposed Echo Park Dam or Bears Ears National Monument are neither simple nor one sided. No matter their opinion, for us, the field is essential in encouraging our students *to have* an opinion about the future of American conservation.

As the trip continues, students become increasingly comfortable sharing their opinions. As we listen during meals, clean up, and sit around the campfire, we hear them express a wide range of views about the use and management of water in Dinosaur, the ongoing development of the oil and gas industry, the status of recreational policies, and funding for infrastructure and amenities within the monument. While there is considerable ground for argument, conversations were never heated. Instead, our students seem more invested in listening and talking to

FIGURE 9.1 Students in the classroom at Harper's Corner Overlook, Dinosaur National Monument.
Source: Andrew McAllister.

FIGURE 9.2 Students in the classroom at the Carnegie Dinosaur Quarry, Dinosaur National Monument.
Source: Andrew McAllister.

one another rather than arguing. Although we work hard to foster this approach through our course design and materials, the field experience itself is instrumental in helping students learn to express and engage with different perspectives without conflict and division. During the trip, they have to work together as they negotiate that space, collaborating and sharing experiences as they set up camp, prepare meals, and complete fieldwork. As we read our students' field journals, we see them applying the knowledge and experience gained from the field trip to position themselves in relation to these landscapes, while also considering and responding to their peers' views and experiences. Field experiences invest students in places and the people they share those places with, and that is why we find it essential to get students out of the classroom and into the wild.

Considering Trip Logistics

Incorporating a course field trip poses significant logistical challenges, which inevitably vary by geographic location, student demographics, and institutional resources. Here we address common challenges and offer suggestions for successful field trips. Given the three-day time requirement, our trip plan coincided with our university's fall break, departing early on Friday (when no classes were held) and returning on Sunday. Since we asked students to attend class during time outside of the official academic calendar, we could not require attendance. We provided advance information to everyone and strongly encouraged participation. In the end, 95 percent of our students attended. Those who did not attend had opportunities to complete local fieldwork assignments with similar objectives to those completed on the field trip.

One of us (Judson), who has extensive experience conducting archaeological fieldwork with students made the design and management of our field trip much easier. Judson's experience encompasses university policies involved with transporting, feeding, housing, and ensuring the safety of students on university-sanctioned trips. We encourage anyone interested in incorporating a field trip into their classes to meet or partner with faculty who have experience planning such trips as part of their curriculum, most often found in anthropology, geology, and outdoor recreation. Instructors should consult with their institution's risk management office to ensure that all official paperwork is completed in relation to safety and liability.[4]

Funding poses another significant challenge for extended trips. We applied for internal grants within our college and departments to cover the food budget, transportation, and other supplies, which allowed us to offer the trip at no cost for students. An alternative to granting is to attach a course fee to fund the trip. Our 2018 trip cost $3,000 (US) and involved 22 students, so a $150 course fee would carry the trip. Though a cost to students may have prevented some from enrolling, $150 is a reasonable cost for a three-day trip that includes food and transportation. Funding opportunities will inevitably vary across institutions, but some combination of grant applications and course fees is a feasible approach.

Partnering with a department that regularly offers field trips as part of their curriculum is a key for supplying essential gear. Judson provided a large kitchen tent, dishes, and coolers to use for the trip, as well as a trailer to transport gear. While many of our students have personal gear (i.e., tents, sleeping bags, camp chairs, etc.), we directed those who didn't to our university's outdoor recreation program, which has a rental shop that offers outdoor gear at a discounted rate. Some students had extra gear that they shared with classmates or partnered up in tents. We rented vehicles from a local rental agency, and, to assist with driving and group management, we invited three experienced archaeology graduate students to join us. Their personal field work experiences made them valuable additions to our group, and this experience, in turn, offered them opportunities to teach and mentor undergraduate students.

Out of the Wild: A Path for Public Engagement

We conclude this essay by coming full circle to our primary pedagogical objective, which is to inspire students to engage the American conservation movement through the public arena of NEPA. From our perspective, students must see themselves as invested in places to embrace a complex civic process in a meaningful way. Our interdisciplinary curriculum is designed to attract the broadest cross-section of students, but our fieldwork is a strategic action to make places like Dinosaur individually relevant. In the opinion of conservationists Gary Machlis and Jon Jarvis, the most successful conservation movements are relevant ones; relevance comes through personal and emotional appeal, and the most potent appeals are to individual and collective connections to nature and history (Machlis and Jarvis 2018, pp. 22–24). This is the very "human environment" that NEPA speaks to in its guiding language.

Following our field trip, we dove into the ongoing debates about Bears Ears, unpacking the complex stakeholder positions and identifying opportunities for public engagement. The 2017 boundary changes under the Trump administration triggered a NEPA review with a complicated and questionable public participation process, and as we introduced students to these issues, they were primed to engage. Drawing on their interdisciplinary studies of Dinosaur and Bears Ears, and motivated by their experiences from the field trip, our students tackled the final group project for the course, producing a best practices manual for public engagement in NEPA review that, in and of itself, was a model for interdisciplinary collective action. This is yet another in a series of strategic actions we see as critical to creating advocacy and stewardship in our student citizen scholars. In keeping with Machlis and Jarvis (2018, p. 48), the future of American conservation requires a long view, where intentional actions like interdisciplinary curricula with complementary field experiences become "innovative experiments in public education … so as to inspire advocacy." The field is the place to connect our students to nature and culture and begin a transfer of knowledge and power to them in a way that matters to their future as well as our own. As other chapters in this volume show, the "field" is broadly defined, but all educators who adapt our approach locally create informed and invested students who are stakeholders in the places that matter to them.

Acknowledgment

Our class was part of the Interdisciplinary Think Tank initiative offered by the Utah State University Honors Program. Feld work was supported by the USU Honors Program, Department of English, Department of Sociology, Social Work, and Anthropology, and the Mountain West Center for Regional Studies. Dinosaur National Monument provided access to cultural sites and camping. We gratefully acknowledge the descendants of the Northwestern Band of Shoshone Indians and the Uintah and Ouray Bands of the Ute Tribe on whose traditional lands we live, work, and teach.

Notes

1 President Biden restored the Bears Ears National Monument boundaries on October 21, 2021.
2 For a link to the handbook our students compiled, see "Playing the Public Lands Game—HONR 3020: Engaging Utah's Public Lands" by Libbie Anderson et al., available at https://digitalcommons.usu.edu/honr_thinktank/7. A description of the assignment can be found on the companion website.
3 See companion website for trip lesson plan and itinerary.
4 See the companion website for examples of trip plans, trip waivers, and safety protocols modeled after the Utah State University Outdoor Recreation Program's forms. These forms have all passed through the USU legal counsel.

References

Advisory Council on Historic Preservation, 2013. NEPA and NHPA: A handbook for integrating NEPA and Section 106. Retrieved from https://www.achp.gov/sites/default/files/2017-02/NEPA_NHPA_Section_106_Handbook_Mar2013_0.pdf (accessed July 29, 2021).

Harvey, M., 1994. *A symbol of wilderness: Echo Park and the American conservation movement.* Albuquerque, NM: University of New Mexico Press.

Jensen, D., 2018. Echoes of opposition. *Journalism History*, 44, 32–39.

King, T., 2013. *Cultural resource laws and practice*, 4th ed. Lanham, MD: AltaMira.

Machlis, G., and Jarvis, J., 2018. *The future of conservation in America: A chart for rough water.* Chicago, IL: University of Chicago Press.

Robinson, R., 2018. *Voices from Bears Ears: Seeking common ground on sacred land.* Tucson, AZ: University of Arizona Press.

Specter, G., 2017. Close looking: Art in the classroom. Retrieved from https://teachingpals.wordpress.com/2017/02/27/close-looking-art-in-the-classroom (accessed July 29, 2021).

Stegner, W., 1955. *This is Dinosaur: Echo Park country and its magic rivers.* Boulder, CO: Knopf Press.

White House, 2016. Proclamation 9558: Establishment of the Bears Ears National Monument. December 28. Retrieved from https://obamawhitehouse.archives.gov/the-press-office/2016/12/28/proclamation-establishment-bears-ears-national-monument (accessed May 9, 2022).

10

CONNECTIONS, RELATIONSHIPS, AND THE LAND

An Anthropology Field School

Kelly M. Branam Macauley, Judson Byrd Finley, Hubert Burdick Two Leggins, Chris Finley and Matthew J. Rowe

American novelist Wallace Stegner famously said, "National parks are the best idea we ever had. Absolutely American, absolutely democratic, they reflect us at our best rather than our worst" (Stegner and Etulain 1983, p. 186). Do they really? For some, sure, but for others, national parks are symbols of colonialism, trauma, and exclusion. Their construction often further displaced and detached Indigenous peoples from their traditional homelands. The National Park Service (NPS) has recognized this and has offered programs and grant funding to engage Indigenous peoples and those underserved by and excluded from our national parks. How do Indigenous peoples engage in these processes to reconnect to their traditional homelands that are now national parks, and how do educators who work to decolonize anthropology and other disciplines aid them in their efforts?

As I packed up my car that afternoon in the early summer of 2005, I was headed to the southern portion of Bighorn Canyon National Recreation Area on the Wyoming–Montana border. This part of Crow/Apsáalooke traditional homeland was mysterious to me. I had lived on the Crow Indian Reservation in south-central Montana on and off for over five years and spent the last year teaching at Little Big Horn College (LBHC) while conducting my dissertation research on Crow constitutionalism. My study highlighted the work of former tribal chairman Robert Yellowtail, who opposed sale of Tribal land for the Yellowtail Dam, ironically named for him and which was deemed essential for providing hydro-electric power across the western US. Construction of the dam near Fort Smith, Montana led to the creation of Bighorn Canyon National Recreation Area, which is bordered by Crow Tribal lands. Even by boat, there is no easy way to get from the Crow reservation to the southern portion of the recreation area. Doing so means at least a three-hour drive through the Bighorn Mountains, into Wyoming, then back into Montana.

When I arrived, two archaeologists I had only recently met greeted me. After introductions to other students, all non-Indian and mostly non-local, the directors

DOI: 10.4324/9781003221807-10

placed me on a team and taught me how to record stone circles. These "stone circles" were the remnant outlines of the tipis that Indigenous people like the Crow once lived in as they traveled the area on seasonal bison hunting trips. They occur in Bighorn Canyon by the thousands along with bison hunting features, sacred sites, and travel markers that are a collective memorial to millennia of Indigenous life. As I sat taking measurements and drawing rocks on my graph paper, I asked the directors, "Where are the Crow people?" All three of us were uncomfortable with the absence of Indigenous people in an otherwise seemingly colonial discipline. We did not want to contribute to the continued exclusion of Crows from this part of their traditional homeland, but how were we to engage Indigenous communities in an academic enterprise like archaeology that has been a historical party to their repression (Thomas 2000)?

This essay describes how we used an ecopedagogy to create a common and shared sense of place that brings Indigenous and non-Indigenous learners into relation with each other and into relation with the land. Anthropology intersects the humanities, social sciences, and natural sciences in a way that is naturally inclusive and intersectional but requires careful cultivation of relationships with Indigenous communities. We follow anthropologist Keith Basso's (1996) notion of place-making as a central concept for ecopedagogies. As Basso notes:

> If place-making is a way of constructing the past, a venerable means of *doing* human history, it is also a way of constructing social tradition, and in the process, personal and social identities. We *are*, in a sense, the place-worlds we imagine.
>
> *(Basso 1996, p. 7)*

We center our project in the simple notion that "America's best idea" dispossessed Indigenous communities of multiple generations from their traditional lands, but an interdisciplinary curriculum that is inclusive of traditional teachings is one way to center "place-making" as essential ecopedagogy.

Our project was founded on relationships of trust and reciprocity with the Crow and Northern Cheyenne Tribal communities, essential principles of Indigenous community collaboration (Kuwanwisiwma 2008). Building trust required forging relationships with individual families and Tribal knowledge keepers; working with Indigenous students of all ages, including K-6 children at a reservation day school and high school seniors in an NSF-funded STEM initiative to increase college enrollment and retention; and, finally, a six-week anthropological field school that provided archaeological workforce training. Reciprocity comes in two forms: first, the teachings of Tribal elders, who, as traditional knowledge keepers, are given an elevated platform along with skills-based learning; and second, equity in the social capital required of Tribal elders to engage their communities balanced with the physical capital (i.e., food, transportation, lodging, and tuition) necessary of any successful field program. At the center of this partnership is the land itself.

This chapter centers the importance of "place-making" in Bighorn Canyon as part of a place-based ecopedagogy. We describe the relationships that were necessary to create successful field programs and how ours became a conduit for Crow people to reconnect to this piece of their traditional homelands, allowing place-making within the landscape to continue. While the federal government has been successful in its centuries-long effort to disrupt these traditional ties, it became apparent that the kind of educational programs we facilitated were perhaps less about Indigenous people "reconnecting" with the land (a somewhat troubling and patronizing perspective) and more about the land reclaiming its people through experiences where an ecopedagogy, traditional or otherwise, has always been central to the idea of place-making. These are potent and lasting experiences for everyone involved, and we share lessons learned along the way that may aid others who partner with Indigenous communities and shorten learning curves in creating successful place-based ecopedagogies.

Historic Removal and Further Displacement by Construction of the Yellowtail Dam

As I sat in that stone circle, to some extent I knew why, as I drove through Bighorn Canyon, stopping at the Devil's Canyon Overlook or one of the four preserved non-Indian historical ranches, there was no mention of the Indigenous lifeways of the Crow Nation or other Indigenous people who lived here for millennia. A detailed observer may recognize or identify the archaeological evidence of these lifeways, but, consciously or unconsciously, the park's informational placards that focused on natural history and the landscape's recreation potential wrote Indigenous people out of the story. The visitor center in Lovell, Wyoming had some materials and scholarly books about Crow Indians, but in the park itself there was no acknowledgment—let alone celebration—that this is Crow land. I believe this is in part because of the NPS ontology, "an implicit organization of the world and its inhabitants" (Trouillot, cited in de la Cadena 2015, p. 76), and the colonial mindset from which it stems. This ontology is often in direct conflict (and has more powerful influence in the construction of national parks) with Crow or other Indigenous people's ontologies of their world, their place in it, and their relationships with the land and other nonhuman occupants.

The history of the Crow Nation, like every other Indigenous nation in the United States, includes removal, displacement, and American acquisition of resources from their traditional homelands. However, my conversations with Crow people about their relationship with the US, one based on friendship and alliance that still resulted in severely diminished landholdings, included thankfulness that their treaties allowed them to remain where the Creator put them: in sight of Awaxaawakússawishe "Extended Mountain," also known as Cloud Peak, "the most sacred place of their world" (Old Horn and McCleary 1995, p. 12). The Crow Nation also retains a small portion of the Bighorn Mountains, the area in and around Black Canyon, where their buffalo herd grazes. From here, one can see Bighorn Canyon from the east. The land the archaeology field school in Bighorn Canyon worked was in sight to the west, but miles away because of the canyon's geography and lack of roads.

Much of the land that makes up Bighorn Canyon was ceded to the United States in 1891. After further land cessions in 1904, the federal government turned its attention to damming the Bighorn River. The negotiations over the land and water rights took nearly half a century and severely splintered the Crow tribe. With a new Crow tribal chairman in 1956, the federal government finally negotiated a price for the land, the dam was built, and the recreation area was established by Congress on October 15, 1966. Having to cross or go around the very large Bighorn Lake to access the canyon and the lands that make up the park's southern unit further displaced Crow people from Bighorn Canyon and made it even more difficult for Crows to visit, encounter, and maintain connection to it. Outside of a few NPS jobs and the occasional fishing trip or individual boat ride, it is difficult to argue that the recreation area benefits Crow individuals or the Crow tribe as clearly as it does non-Indian boaters, fly fisherman, campers, and the multi-billion-dollar businesses that support these activities.

We wanted to assist in changing this. Could the recreation area benefit the Crow tribe? As the NPS was addressing the harm it has caused to Indigenous communities, working collaboratively with NPS leadership and Crow elders, we felt we could at the very least bring Crow people into Bighorn Canyon through archaeological educational and training opportunities. Our first major obstacle was community trust and buy-in. How could we get Crow people to travel several hours away from their homes for a single day, let alone convince them that a six-week educational camping trip would be a good idea? We knew it would take time to create this field school, and our approach eventually led us to be in reciprocity with each other, the land, and the environment in new and different ways.

Building Relationships

The relationships required for a successful collaboration with Indigenous communities begin at the individual level. They take time to build. I had become a part of the Crow community, and even when I left for extended periods, I remained in touch and was expected to return. These relationships are essential for an Anthropology field school like the one we envisioned. The best approach is to partner with someone who has done so, always keeping in mind that the best relationships are built on reciprocity. During my time at Crow, there were a few women who had taken me as "kin," their sister, daughter, or granddaughter, and, as I could, I brought them and anyone else to visit the field school. We learned from these visits that time, money, camping supplies, and child-care made visiting Bighorn Canyon difficult. My family members would have to take off work, have reliable transportation through the mountains, buy or borrow camping supplies, and our camp of college students needed to be kid friendly. While most families own tipis, camping tents were not often found in Crow households.

Getting Crow people to Bighorn Canyon was proving difficult, so to build relationships we planned field trips for the archaeology field school students on the Crow Reservation with LBHC colleague and anthropologist, Tim McCleary. We formed a relationship with the Pretty Eagle Elementary School in rural St. Xavier, Montana, which the McCleary children attended. In the evenings, the school's community room became a presentation space, where tribal educators spoke about Crow culture, and an eating hall where we shared food and visited with friends and former students. Eventually, the school administrators invited us to "camp" in the community room and share breakfast with their students. In return, working with McCleary, we organized day trips for Pretty Eagle students and archaeology field school students to local archaeological sites in the Fort Smith and Bighorn Canyon area that provided an opportunity to center traditional place-based oral history in archaeological education. This provided the foundation of a trust relationship with Crow educators that, with time, provided the community support necessary to make the program we envisioned successful. But we wanted Crow students, educators, and elders in Bighorn Canyon with us, and we needed to build relationships that would bring them there.

On a four-day break from a ten-day session of the archaeology field school in summer 2008, I was sitting in a café on the Crow reservation. One of my former students, Noel Two Leggins, walked in, recognized me, and asked what I was doing and why I was back. I told him about the archaeology field school, our visit to Pretty Eagle Elementary School, and our joint trip to Plenty Coups State Park, also on the reservation. He said I should talk to his dad, Burdick, and tell him about our work in the Canyon.

Several days later, Hurbert "Burdick" Two Leggins and several of his children visited us in Bighorn Canyon. Burdick was a member of the Crow Culture Committee, a group of Tribal knowledge keepers who provide oversight of wide-ranging activities including advising the then newly formed Tribal Historic Preservation Office (THPO), a federally designated participant in legislatively mandated environmental review processes related to the development of Tribal energy resources like coal and natural gas. This turned out to be one of many visits, and our friendship and partnership created the shared vision of the field school that would still take four years to build. As we visited a former tipi encampment the archaeology students worked on, Burdick gathered his family around the perimeter of a stone circle and said, "Respect the rocks and the circles. They are sacred. Do not take the rocks away. If you take one of the rocks away, they will be lonely. So, leave the rocks where they are." This was a stark reminder that archaeological sites are places of memory for many Indigenous people. The message was one of cultural preservation, and the implication was that memory and preservation could be passed down through education. Archaeology suddenly became relevant as an educational tool where the land holds memories that link past, present, and future in a timeless circle.

The next day we visited the Bad Pass Trail, an ancient travel route marked by rock cairns. As Burdick was talking, we saw on a hill across from us seven bighorn sheep. I thought to myself, "It is as if this place, this environment, and the animals behave differently when Crows are around." That evening, Burdick shared the Crow story of Big Metal. In that story, a young boy is thrown over the edge of

Bighorn Canyon by his stepfather. The boy is caught in a juniper tree on a ledge of the cliff face where he is rescued by *seven bighorn rams*. The leader of the rams, Big Metal, bestowed the boy with his own name, asked other animals to give the boy gifts, and shared other important knowledge. Burdick's story was a reminder of people being in relation with each other and the land where they live. He reminded his children and grandchildren that they have three mothers: their biological mother, their tipi homes (much like the womb from which we all come), and the Earth itself. This land, the canyon, and the river was to forever be called Bighorn in remembrance of Big Metal, the sheep. Big Metal, the man, lived for many years and was buried somewhere between St. Xavier and Fort Smith where we had been with the archaeology field school students just weeks before.

As this story, and the place-making that comes from its telling, shows us, the Bighorn Canyon landscape transcends these moments in time. Remaining on the ground are the stone circles from a time before the Crow people were given tipi stakes when Big Metal (ram and man) walked *in the canyon*, and we are still listening to these stories there today. The stories instruct Crows how to live. Place-making is important to Crow identity; Crows are the "place-world they imagine." They are the "chosen people," chosen by the First Maker to live in "just the right place" and care for the sacred tobacco seeds found at the foot of the Bighorn Mountains (Branam 2008). Being in Bighorn Canyon and telling these stories continues this place-making tradition, grounding the Canyon, the bighorn sheep, and everything else in it within the identity of Crow people. As Basso (1996, p. 34) reminds us, "Knowledge of places is therefore closely linked to knowledge of the self, to grasping one's position in the larger scheme of things, including one's own community, and to securing a confident sense of who one is as a person." Burdick's visit showed us just how important this land remained to Crow people even though they had been disenfranchised from it through more than a century of federal policy. Furthermore, we realized the potential of archaeology field methods to serve as an educational conduit in ongoing Crow place-making that could bridge generations with elders passing knowledge through Indigenous and non-Indigenous students alike. With Burdick's shared vision of our program, we were able to create the community partnerships necessary for a truly collaborative field experience.

America's Best Idea: Traditional Knowledge and STEM Education in Bighorn Canyon

Later in summer 2008 I was back in a car driving out of Bighorn Canyon with Chris Finley, the recreation area archaeologist, who asked me to come with him to Greybull, Wyoming to see an amateur historian's presentation about the town of Greybull to descendants of the Crow Chief Greybull. On this day we were introduced to Mary Bear Cloud, several of her adult children, and a few of her grandchildren. As it turned out, I knew one of her daughters from my time at LBHC. We talked to Mary and her family about the archaeology that was taking place in

Bighorn Canyon. Mary knew of the sites there and now learned that white archaeologists were working on them. She was concerned by our work, wanted to get to know us and our intentions better, and soon visited Bighorn Canyon bringing along her adult sons. She understood our meeting as an opportunity to use her influence as a traditional knowledge keeper to steer our direction in terms of the archaeology we were conducting, the field school we were creating, and for her family and other Crows to reconnect to this area of the Canyon in their traditional homeland.

As the NPS prepared for its centennial celebration, and in advance of the release of Ken Burns's 2009 PBS documentary, "The National Parks: America's Best Idea," the non-profit National Park Foundation piloted a grant program that tackled head-on the central notion of Burns's documentary based on Wallace Stegner's infamous quote that leads this essay. The intent of the grant program was to provide educational opportunities, particularly among underserved and Indigenous communities, to connect (or reconnect) with America's national parks in meaningful ways. We saw an immediate opportunity to meet a central part of our agenda of bringing Crow tribal members to Bighorn Canyon for an extended stay, in this case high school juniors and seniors participating in a project through LBHC to foster STEM education in Tribal youth and increase college enrollment and retention rates. Our approach paired traditional place-based education provided by Tribal knowledge keepers with the STEM applications of archaeological field methods. Our vision was a five-day camping trip at Bighorn Canyon that used the infrastructure of the archaeological field camp—equipment and

FIGURE 10.1 Setting up a tipi in Bighorn Canyon National Recreation Area. Source: Northwest College, Powell, Wyoming.

student assistant instructors—as its base. The Foundation agreed that this was a good idea and supported the project for 2009 and 2010.

As Mary Bear Cloud was fond of saying, "Everything happens for a reason, and there are no coincidences." The Crow students who participated in our project were accompanied by several chaperones, one of which was Mary's daughter who I had known from my work at LBHC and who had visited Bighorn Canyon earlier that summer with her mother and many other family members. Burdick Two Leggins likewise used his influence as a member of the Crow Cultural Committee to support Tribal investment in the opportunity. The trust and reciprocity that we knew as essential ingredients to a successful collaborative relationship were taking hold. In hindsight, the STEM-based curriculum is less relevant to us than the experience that the opportunity afforded as the land began to reclaim its people from youth to elder. Yes, students learned principles of archaeological surveying, cartography, and site documentation, but more importantly, Tribal knowledge keepers, Mary Bear Cloud among them, shared with all participants the teachings they deemed most important and in ways they chose to deliver them—often spoken in the Apsáalooke language. Bighorn Canyon saw its first Crow tipi encampment in perhaps more than 150 years, and the landscape echoed with drumbeats and traditional song. This act of facilitating people in a place through a STEM-based ecopedagogy helped return them to being in relation with this land. The success of the "America's Best Idea" project paved the way for the full six-week anthropology field school that was to come.

Archaeology as Ecopedagogy in Tribal Workforce Development

A standard university-based archaeological field school is a six-week, six-credit-hour field class that often requires travel to remote field locations. Field sessions are three consecutive 10-day hitches with two 4-day breaks in between. Students camp and work under all conditions and meals are prepared communally. Camp cleanliness, including food storage and bathroom facilities, is a major logistical challenge to maintaining student's physical and mental well-being. Bighorn Canyon is also prime bear habitat, which adds an extra dimension for managing the archaeological camp. These details, along with the research and educational components, are part of the instructors' responsibilities. Yet this kind of training, along with a bachelor's degree in anthropology or closely related field, is considered essential for entry-level positions as a professional archaeologist. Archaeological field schools are also challenging for most traditional college students because of tuition and field fee costs (usually a set rate added to tuition), the requirement of being away from home, family, and a source of income for six weeks, as well as the cost of essential gear necessary to be in the field. These issues are amplified in Indigenous communities making this kind of educational opportunity even more difficult to realize.

By 2010, the Crow and neighboring Northern Cheyenne Tribal Historic Preservation Officers (THPOs) faced the reality that archaeological training was an essential part of the workforce development they needed to provide skilled Tribal monitors on archaeological projects. These monitors worked alongside non-Indigenous archaeologists, who had the training we describe, at archaeology projects on Tribal lands. Without equivalent training, many Tribal monitors were disadvantaged on professional crews. THPOs from across the northern Plains Tribes were concerned about the lack of training and the credibility of their archaeology programs. To address this need, our existing archaeological field school, combined with NPS grant support and the relationships we had cultivated with the Crow community during previous years, provided the framework for a successful THPO training program launched in 2011 and 2012.

We knew from the beginning that for a Tribal archaeology field school to operate successfully, our biggest challenge would be recruiting and retaining Indigenous students. Asking Crow and Northern Cheyenne students to attend an archaeological field school was one thing; getting them to stay for 10 days was another; convincing them to return after a four-day break was yet another. Repeating all of that for successful completion of the three 10-day sessions was the ultimate challenge. We were successful because of the time we had taken over the previous years to develop strong relationships with tribal partners based on trust and reciprocity. Burdick Two Leggins, Mary Bear Cloud, and other Tribal knowledge keepers used their influence, primarily within their extended families, to recruit and retain students, who ranged in age from 18 to 54 years old. We hosted entire

FIGURE 10.2 Students learn to make and identify stone tools in Bighorn Canyon National Recreation Area.
Source: Northwest College, Powell, Wyoming.

families—spouses, children, and grandchildren alike—in our camp, providing food and educational opportunities for everyone. We hired Burdick's spouse, an accomplished camp cook, to make sure everyone was fed and happy. Her daughter and granddaughter assisted in the field kitchen. In our experience, a happy camp is half of the ingredients for a successful field school. We found those ingredients through including entire families into our camp.

The other part of the recipe that made this project successful was an inclusive curriculum. While archaeological workforce training was the objective, based on our previous experience, that curriculum had to be expanded beyond traditional skills-based learning to give equal footing to Tribal knowledge keepers who visited our camp as guest lecturers. Knowledge keepers spoke about topics of their choosing, including the importance of maintaining traditional language to effectively communicate one's heritage, place-based oral traditions that directly linked Tribal history to the Bighorn Canyon landscape, the vital role those Tribal members play in preservation of cultural heritage through a participatory process mandated by the federal government, and messages about the importance of strong family bonds and ties to the land. Each of these messages, and their carrier, provided yet another reason for the students to buy-in to the process that we knew would be critical to the program's success from the very beginning. Our Tribal collaborators were able to close the circle of a place-based ecopedagogy founded on traditional knowledge and archaeological method that helped bring Indigenous and non-Indigenous students[1] into relation with the land in a way that we alone were not capable of doing. The "slow learning" that unfolds over the course of six weeks creates the lasting sense of presence with the land and community among participants that realizes Basso's (1996, p. 34) ideal that knowledge of place is knowledge of self. We think this is the real power of a collaborative community-based ecopedagogy grounded in the vision of America's best idea—its public lands and national parks.

Practical Matters to Consider

We describe three kinds of field-learning opportunities, each of which requires different levels of participant investment. One way to enact an ecopedagogy that is place-based and centers traditional knowledge is to meet students where they are at. Partnering with local K-6 schools and traditional knowledge keepers is one way to accomplish that (for an example, see Maier this volume). Programs like the week-long STEM field school and six-week field school that bring Indigenous and non-Indigenous students to a remote field location where they learn archaeological methods and interact and learn from Tribal elders takes considerable funding and coordination. Grants supported scholarships for tuition and field fees for Tribal college students, as well as salaries for support staff, including teaching assistants and camp cooks. Sometimes it is necessary to provide camping equipment for Indigenous students and guests, so a budget for equipment is also important. An inclusive camp that brings entire families can quickly expand from 15 students to

30 or 35 people, and sometimes more when guests are counted. The food and water demand of a camp that size is extensive. While the NPS provided a 250-gallon water buffalo, we handled the logistics of meal planning and shopping ourselves. NPS vault toilets were available for our use, otherwise portable toilets and their regular maintenance are an essential cost. We were fortunate to have a freshwater stream at camp and the reservoir a short distance away for bathing, otherwise solar showers are an important part of a good camp. The logistics of these essentials is very important in keeping a happy camp, and comfortable students are better learners.

Financial support for Tribal knowledge keepers, who are often elders, is another major consideration. These elders may not drive and require a traveling companion. Their needs may make an overnight camping trip impossible or a day trip very difficult. We learned that for an elder to visit us meant that their son or daughter would have to take a day off work to bring them. Understanding this, we wrote "care-giver" stipends into our grants. If an elder could not make the entire trip, we sometimes incorporated their visit into a field trip, meeting them at a site like the Bighorn Medicine Wheel, cutting the drive from the reservation in half. Elders received an honorarium in the form of cash and a set of gifts that remain part of traditional knowledge exchanges in many Indigenous communities. These are all essential considerations in project budgets that are critical to the kind of trust and reciprocity necessary for successful inclusion of traditional knowledge in place-based ecopedagogies.

We know through experience that a lot of life happens during field projects like the ones we have described. Animal encounters are real. Bears come and go and sometimes prefer to stay, wreaking havoc on canvas kitchen tents saturated with the smell of bacon grease. Everyone must know proper bear country protocol about food and personal items in tents. Thunderstorms and other extreme weather events pose real risks for archaeological fieldwork that finds students hiking miles from camp each day. In remote locations with no cell phone or internet reception, instructors must be ready for all scenarios. Students, Indigenous and non-Indigenous alike, have difficulties adapting to the loss of cell phones and internet to the point of obsession about daily drives to the closest point they know for reception. Sometimes this is a practical matter of checking in on family, but we find it fades with time. We filled downtime in camp with traditional crafts like beading, flintknapping, leatherwork, yoga sessions, and even hatchet throwing competitions. Idle hands in a remote field camp are truly problematic and filling down time with quality experiences only adds to the sense of community that unfolds naturally in the field, however we choose to define it.

In closing, we note that there may be a spiritual risk that Indigenous students take by working with archaeological sites. It is important to be sensitive to who can interact with what kinds of sites and in what ways. Burdick Two Leggins, as a key leader of our team, attended to the spiritual well-being of everyone who participated in the field school, performing smudging ceremonies as needed and developing protocol for the proper treatment of sites that included offering tobacco to honor the ancestors who had lived there and left the sites we recorded. These kinds of inclusive practices for us only highlight the centrality of diversity, equity, and inclusion in a successful place-based ecopedagogy that actively engages Indigenous communities.

Final Thoughts

As we got up and resumed our surveying of the Bad Pass Trail after lunch in 2012, I looked back and saw Burdick stop, pause for a moment, and place a rock on a cairn. I had heard about this practice before but had never seen a Crow person do so, asking for good travel and for our safe return to this place. I felt so fortunate to witness this small, powerful act of Crow place-making continue in Bighorn Canyon. Crow ontology includes nonhuman agency, and we are reminded of that when they tell stories like the one of Big Metal and can restore their connection to the land through their own place-making. For non-Indians, any acknowledgment or partial understanding of Crow ontology may only be made possible by being in Bighorn Canyon itself, being in the landscape with Crow elders. To be in relationship, Indigenous and non-Indigenous people of all ages must go into the field together. The field school we finally realized was a conduit of reconnection for Crow people and for them to share their place-making with us. In doing so, they made us better anthropologists. Placed-based education *is* place-making. It is how we use ecopedagogies to create a shared sense of place that brings people into relation with each other and into relation with the land.

Acknowledgment

Fieldwork for the projects we describe was made possible by the US National Park Service Bighorn Canyon National Recreation Area, the National Parks Foundation, and Northwest College (Powell, Wyoming). We are indebted to the many community members and students who helped us along the way. Alexandra Hedquist read and commented on early versions of this draft. We are grateful to Dr. Elizabeth Baldwin and Dr. Maureen Boyle for their support in and out of the field. This work was conducted on the traditional lands of the Crow and Shoshone Tribes of Indians, who remain stewards of the land today.

Note

1 The 2011 field school students were all Crow and Northern Cheyenne Tribal members. At the end of the season, several students commented that they would have enjoyed the experience more with students from other schools. In 2012 we combined the field school to include both Indigenous and non-Indigenous students.

References

Basso, K., 1996. *Wisdom sits in places*. Albuquerque, NM: University of New Mexico Press.

Branam, K., 2008. Constitution-making: Law, power, and kinship in Crow Country. PhD thesis, Indiana University.

de la Cadena, M., 2015. *Earth beings: Ecologies of practice across Andean worlds*. Durham, NC: Duke University Press.
Kuwanwisiwma, L., 2008. Collaboration means equality, respect, and reciprocity: A conversation about archaeology and the Hopi Tribe. In C. Colwell-Chathaphonh and T. J. Ferguson, eds. *Collaboration in archaeological practice: Engaging descendant communities*. Plymouth: AltaMira Press, 151–170.
Old Horn, D., and McCleary, T., 1995. *Apsáalooke social organization*. Crow Agency, MT: Little Bighorn College.
Stegner, W., and Etulain, R., 1983. *Stegner: Conversations on literature and history*. Salt Lake City, UT: University of Utah Press.
Thomas, D., 2000. *Skull wars: Kennewick man, archaeology, and the battle for Native American identity*. New York: Basic Books.

11

LEARNING TO THINK LIKE A FACTORY

Lachell Faure, Darren Paisley and Brad Tabas

On Machines and Gardens

In his 1964 classic *The Machine in the Garden*, MIT professor Leo Marx explored the ways in which the symbols of the machine and the garden crystallized the conflicts between industrialization and its arcadian or environmentalist enemies over the course of the nineteenth and twentieth centuries. His work was retrospective, yet also prescient. Even today debates around technology and ecology tend to embrace one of these two symbolic poles. Some advocates of the machine, or, to borrow a term from Daniel Deudney (2020), "Promethean Technophiles," claim that the only solution to problems like climate change is technological innovation, including new sorts of energy and even geo-engineering technologies. "Luddite Technophiles" (to again quote Deudney) take the opposite tack, embracing a return to the garden in the form of a future that they imagine as low-tech or semi-primitive. It seems that Marx himself, whose work can be seen as one of the earliest attempts at what today would be known as eco-criticism, and whose courses at MIT should probably be acknowledged as forerunners of our current attempts to integrate environmental responsibility within engineering education, wanted to embrace neither pole of this rigid opposition: he wanted to explore ways in which both the machine and the garden could exist in harmony, ways in which *the* machine could be *in the* garden.

Like Marx, we teach at an engineering school, and, like him, we have been inspired by the idea that at some point in the future our high-tech industrial civilization will be able to reconcile with the garden world on which we dwell. We have thought about ways in which engineering education can contribute to making this possible. We explain part of our answer in the following introduction to a class that we collectively teach to our engineering students called the "Circular Economy Project." A circular economy, according to the definition given by

DOI: 10.4324/9781003221807-11

Christian Arnsperger and Dominique Bourg, is one in which our material flows have been comprehensively and exhaustively minimized such that what goes in to, and comes out of, our production and consumption modes is as close to zero as possible (Arnsperger and Bourg 2017). This idea, or rather this paradigmatic ideal for the future of the human economy, dates back at least to the work of Barry Commoner and early 1970s thinkers first reacting to the discovery that there are planetary limits to growth, though, as Robin Wall Kimmerer and Laura Harjo remind us, First Nations people have long embraced an ideal of circularity in their economic and social practices.

This year-long project includes periods of classroom study as well as periods of in-company work experience research aimed at helping our students to develop a sense of, and an appreciation for, the healthy factory—the kind of machine that might co-exist with the garden as opposed destroying it. Our students are apprentice engineers already actively engaged in industry. As a result of our project, we particularly wanted to raise the student's awareness of where stuff comes from and where it goes, information that at the outset they were little aware of even with respect to their own companies and industries. The Circular Economy Project prompts the students to find out—as far as possible—about the situation pertaining to their companies in terms of economic circularity or linearity. The Circular Economy Project encourages them to learn about how the companies in which they are apprentices, as well as those of their peers, are striving to make their production as ecological and sustainable as possible by investigating how they consider—or don't—the upstream and downstream environmental and social effects of their production. The other objectives of the project include helping students to acquire industry-specific know-how concerning environmentally sensitive best practices and ecological awareness and concern regarding how their future professional lives may impact their local and global biotic communities. Playing on a phrase that Aldo Leopold (1966) used to express the essence of his land ethic—learning to "think like a mountain"—we describe this as learning to think like a factory.

When one thinks like a factory, one takes a holistic approach to thinking about the nature of industrial production. Drawing on the work of environmentally engaged economists such as Kate Raworth (2017), we prompt our students to see that a healthy factory, from a certain point of view, looks something like a circle or donut. To get to this abstract donut in which inputs and outputs are revalued requires other attunements. As Steven Vogel (2015, p. 129) puts it, in a book where he encourages us to learn to think like a mall, this kind of thinking requires tangling with "the dark complexity and depth of the processes of nature" and the fact that our simple desire to master these objects, to "grasp and control" them, just won't work. Of course, this recognition of our inability to master the situation does not imply that we should renounce any responsibility to this complexity, but only that we need to engage with these things relationally, listening to and caring about them. Carole Gilligan (2016), Nell Noddings (2013), Maria Puig de Bellacasa (2017), and Virginia Held (2005) have all theorized ethics of care, moral philosophies emphasizing the importance of relations and relationality, including our relationships to ecosystems and social systems, both human and extra-human. A primary aim of the Circular Economy Project is fostering in our students a sense of care regarding their workplaces and the larger world. Developing an ability to care for factories is anything but obvious: human beings find it easy to care about charming things like their own children and scenic woodlands, but they find it far harder to care about the snot-nosed urchins in the street and industrial parks. It is in many ways more important to care about things like shopping malls and factories than the ecology of beautiful wild mountains, if only because it is these kinds of places that make up most of our world. When a mall or a factory is truly healthy, it can be part of the flourishing of a larger biotic community, a node in a successful entanglement of nature, technology, and culture. When it isn't, it becomes an eyesore. It produces alienated and dissolute workers who stain the nearby communities with their misery. It belches out pollutants into the nearby air and water, it produces products that in their fabrication rape the Earth, in their utilization turn our blue skies brown, and in their afterlives flood our landfills and spill out into our rivers. Uncared for factories inspire environmentally careless communities, while encounters with well-tended industries can inspire further engagements and actions. To return to our metaphor of the machine and the garden, we might say that the hallmark of a carefully designed, used, and disposed of technology is that it is able to co-exist with—even embellish—a garden, and that the art of learning to think like a factory involves learning a sense of how to make this the case in a way that goes deeper than greenwashing.

The Circular Economy Project

The Circular Economy Project, which might be categorized as what a recent MIT report (Graham 2018) calls "work-based learning," includes a series of classes, fieldwork, and reporting activities and is taught to teams of apprentice engineers—engineering students who alternate between periods at our institution and periods in company. This is important to note, because the pedagogical strategy we employ requires that the students be able to observe and investigate the conditions in place in-company (this is possible for students completing internships as well). It is also important to emphasize that this is a project for teams of students and that a large part of the learning involved in this project—particularly with respect to its ethical dimensions—derives from exchanges between students regarding their differing professional experiences. It is also worth noting that the Circular Economy module was part of our student's English language curriculum. All courses were taught in English with English-language readings and videos, and all discussions and reports were given in English, though most (but not all) of the students' employment experiences were in a French-language context. The rationales for the inclusion of this module within the English curriculum are multiple. On the one hand, and from our point of view as ESL teachers, encouraging advanced (B2+) students to do research and make reports in English was a highly effective way of encouraging them to learn new vocabulary and to develop high-level speaking and writing skills. Fostering environmental ethics in an ESL classroom was also a means of fitting additional environmental responsibility-oriented pedagogy within what was already a very tightly packed curriculum.

The Circular Economy module is split into three distinct sub-sections followed by a final, summative assessment after the last class. Part 1 takes place during students' first academic sequence at the school. To put this in context, they will already have completed their first professional sequence (they will have already begun to work at the company which will employ them as an apprentice engineer over the next three years.) Following two months at school they then return to their companies, but while at work they continue doing research for the Circular Economy Project and communicating with their teammates. Two months later they return to the school for the next academic sequence and the final stages of the project, which culminates in a group presentation before a jury. The module is therefore stretched out over a period of around seven months with differences in time and approach marking each of the three stages.

Across all sub-sections we encourage students to make an honest, clear-sighted, and non-judgmental assessment of the ways in which notions such as recycling and recyclability, product lifecycles, circularity vs linearity, and so on are present or absent in their companies. It is important to note that our aim is not to vilify their place of work or to provoke conflict between students and their companies but to make them aware of the rewards and the challenges confronting efforts to make our industrial economy more sustainable.

If the class is in many ways about immersing students in the realities of the Circular Economy, we nevertheless begin the module in classrooms and with concepts. Overall, (and expressed in terms derived from Bloom's taxonomy), the different phases of the class are devoted to accomplishing the following learning objectives:

- Part 1 (in school) involves *identifying* and *understanding* key concepts (as well as sub-group organization and planning).
- Part 2 (in company) sees the students *applying* the concepts to their workplaces in a fact-gathering exercise, with them subsequently *analyzing* the relevant type(s) of procedures in their respective companies. This stage also involves a certain amount of synthesis as the sub-group members share their findings with each other and coordinate their activity.
- Part 3 (in school) starts with the sub-groups of students *synthesizing* their findings (a process begun in part 2) as they bring together their information, experience, and ideas. The sub-groups then build on this by *evaluating* and *contextualizing* their work.

This is obviously a very broad picture since many of these steps operate in a parallel or otherwise non-linear fashion. We will now briefly look at each part in a bit more detail.

Part 1

What Is the Circular Economy?

This part is made up of four two-hour-long classroom sessions which bookend students' time at the school. In the first classes of Part 1, we introduce students to the topic and divide into sub-groups. Starting early allows us to be sure that group members have time to work together before they head off to their individual companies, and it also ensures that they will have easy access to teachers who can answer questions regarding any concepts that they find difficult. Likewise, concluding their semester with a class allows us to be sure that students are on-track before they enter observation mode upon returning to their companies.

The first class introduces students to the module, and it also offers a teacher-centered introduction to the theory of the Circular Economy, which, depending upon the teacher involved, means slideshows, short films, questionnaires, and discussion sessions. Students learn about topics such as recycling, lifecycle thinking, and planned obsolescence. This introductory class is also used to divide the students into the sub-groups in which they will work until the end of the module. In addition, we also try to involve outside practitioners during this phase. The school is associated with a small-business incubator which includes several start-ups actively involved in different aspects of the Circular Economy. We ask one or two of these entrepreneurs to come to the class and present their activities to the students. The students welcome these guest talks, and the feedback is uniformly positive and enthusiastic. At the end of Part 1, we ask each group

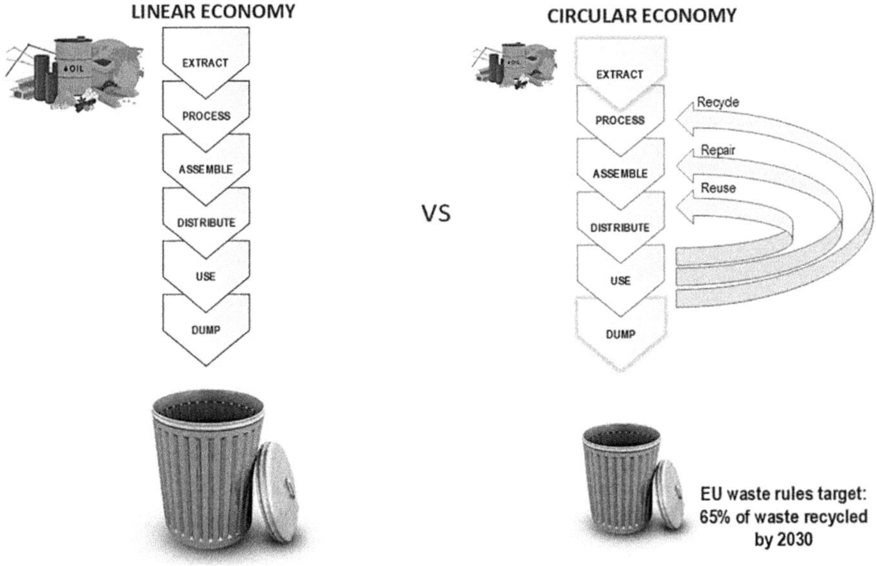

FIGURE 11.1 Comparison of Linear versus Circular Economy.

to give a short talk detailing their understanding of the Circular Economy. From the teacher's point of view, this talk constitutes the first "learning milestone" and the first possible evaluation. From a pedagogical point of view, the primary aim of these talks is to invite students to learn how to work with their team members as well as to encourage them, by working together, to deepen their understanding of the notion of the Circular Economy. In the same session teachers also brief the students on their expectations regarding the work to be done in-company. Students are asked to stick to the facts. The object is neither to guild the company nor to put it on trial. The object is to give as honest and complete an account of the reality in-company regarding the Circular Economy. The object is also to focus on the activity of the company. We emphasize, for example, that they need to attend to what happens with a company's primary products and not what happens regarding the recycling of the disposable coffee cups in the cafeteria.

Part 2

Looking at the State of the Circular Economy in Specific Companies

During this phase, which could be characterized as information gathering and analysis, the students are dispersed across factories, shipyards, and design offices throughout France in their capacity as apprentice engineers. They are isolated from one another, yet they must continue to work together on the module requirements. We encourage students to keep a simple logbook briefly detailing—from day to day—their observations. They can then analyze the entries at their leisure, decide what needs to be pursued, and choose what they will share with the other members of their sub-group. Again, we encourage students to adopt an honest approach. They find out what they can, whenever they can, looking sometimes at the company's published statements and sometimes at the day-to-day practices that they see going on around them at work. If there are barriers to understanding (official, practical, etc.), they duly make a note of these. At the heart of this sequence is the attunement of their ability to observe and note what goes on in the professional world.

Part 3

Processing

This is the final phase of the module, consisting of four two-hour sessions which once again bookend their time at the school. In the first session, groups present a clear, factual account of how far their companies are (or are not) engaged in the Circular Economy. This report is informal, and students are meant to understand it as part of their learning process, with class discussions regarding the differences between companies' practices being one of the aims of the exercise. The same information—only more carefully presented and accompanied by a more detailed commentary and analysis—will form the basis of the final, formal presentation before a jury. The three subsequent classes aim at encouraging students to move from their specific findings and experiences to more general observations regarding the state of industrial production in France and elsewhere.

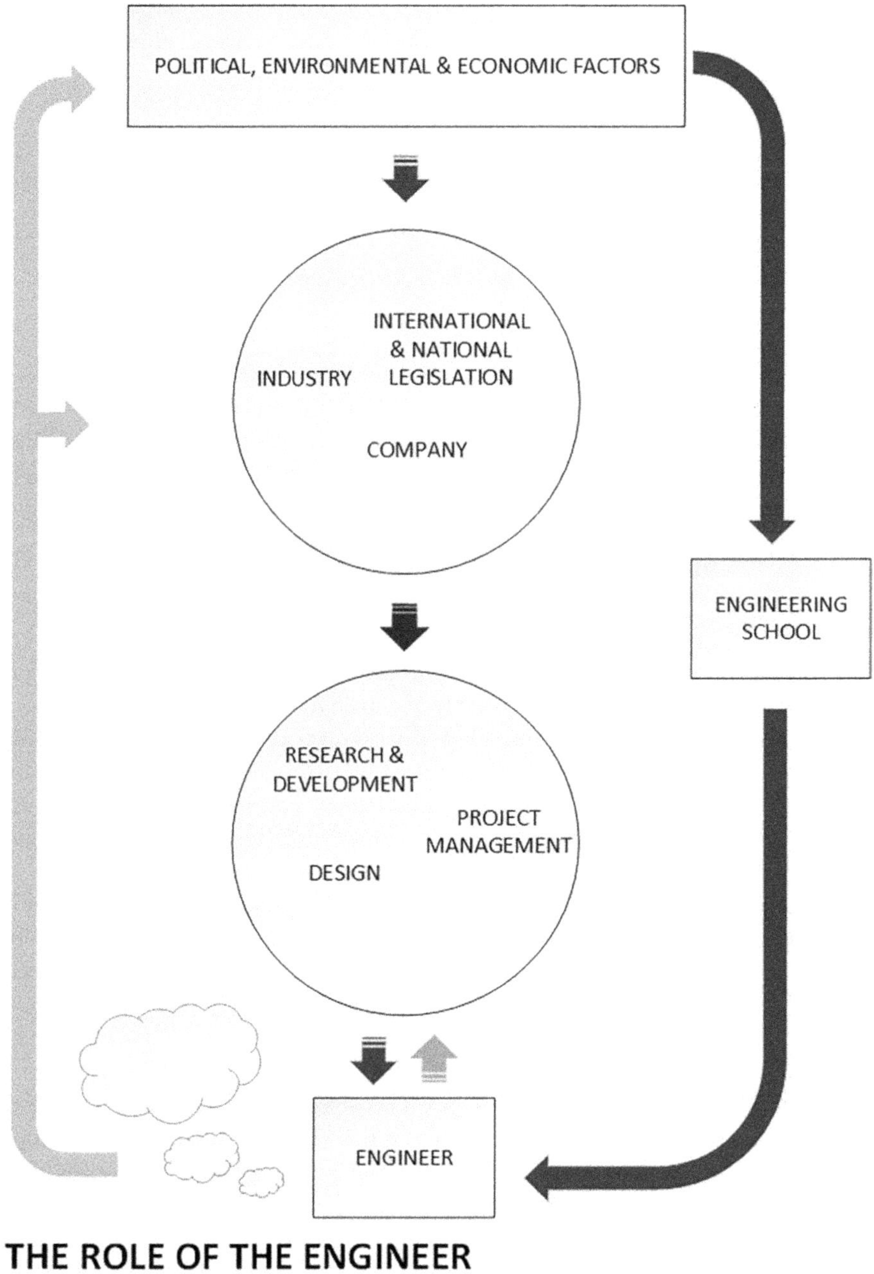

FIGURE 11.2 The Engineer's Role in a Circular Economy.

We encourage students to think about their respective roles within industrial production and to think about how future engineers might endeavor to make their companies more sustainable (i.e., circular). To encourage this kind of thinking, we prompt them to reflect on questions such as:

- Does current engineering training equip students to work in or towards a Circular Economy?
- What might be done to improve the situation?
- What are the barriers blocking such an improvement?
- How could improvements be achieved, and what role would the engineer play in implementing these?
- Are there models to follow?
- What future is desirable?
- Which future is possible?

To sum up this period, teachers ask students to bring their findings together in a coherent manner, evaluating the engineer's role in the articulation of current and future industry in light of the Circular Economy—with a firm distinction kept between the actual, the possible, and the desirable. They are thereby also encouraged to develop a clear and experienced-based sense of the various ways in which companies are responding to the challenge of becoming more sustainable.

The Group Presentation

The final session consists of a series of group presentations that last 20 minutes and include a 15-minute Q&A session. Arguably, the main function of these presentations is to develop communication capacities, skills that are generally important for engineers but which have also been highlighted as crucial by many studies done regarding the key-competences for training sustainable development engineers (Wiek et al. 2011; Mukhtar et al. 2019; Quelhas et al. 2019; Shahidul 2020). The presentations also prompt students to further reflect upon what they have seen and learned. We try to encourage jury members to ask questions bearing on the way forward, asking them to point out gains that might be realistically accrued in their workplace, perhaps by applying practices that exist in other companies and on which their colleagues reported.

Feedback

At the end of the school year the 41 students involved in this project were asked to complete a survey bearing on how the courses that they had taken over the course of the year impacted their levels of environmental awareness and engagement. This was a qualitative survey involving three short-answer questions:

1. Have your classes at the ENSTA made you more aware of the responsibilities of engineers towards the environment?

2. Which classes and which exercises have most contributed to raising your awareness about the responsibilities of the engineer?
3. It is easy to translate what one learns at the ENSTA about environmentally responsible engineering into one's professional career. How strongly do you agree with this statement? Explain your answer using real examples if possible.

Note that this survey did not bear only on the Circular Economy module but on the totality of the students' first-year course work. The results suggest that the students found the Circular Economy module to be a particularly valuable part of their educational experience.[1]

Question 1

Nearly all the students agreed that they had learned more about the environmental responsibilities of the engineer over the course of their semester. In fact, only 2 out of 36 claimed to have acquired no awareness, but those that made these claims offered no reproach to the school or the curriculum. As one student explained: "I chose such studies in order to work in these matters [environmental engineering] so I was quite aware before." Another student claimed, "to have already been quite aware upon arriving at the school" but nevertheless highlighted that studying environmental responsibility in an English class "was a good way of deepening our awareness." Two students acknowledged that they had become more aware of these matters while at the school, but they regretted that they were not more emphasized, with one student alleging that it was "only in English" and particularly in the "Circular Economy" module that environmental issues were discussed. Among the 32 merely positive responses, appreciations ranged from the non-committal "yes, a little" to what we, as teachers, would categorize as an ideal reaction to this pedagogy:

> Yes, of course. Especially the work on Circular Economy. If I had not done this research on my company, I would never have known that there was a service that developed tools to measure the environmental impact of our products. And I've been working in my company for 2 years! Thanks to this research, I feel more concerned about this aspect. And now I am registered on the environmental intranet group of my company.

Question 2

Twenty-five students specifically cited the Circular Economy module as the one that contributed the most to expanding their environmental awareness. As one student put it, "This project allowed us to focus our research on 'what our company really does to be sustainable.' It was a really interesting exercise!" Seven students somewhat more broadly but also more vaguely emphasized that it was their English classes that contributed most to developing their awareness and sense of responsibility, with some students specifying that it was another course, "Science and Society" (also taught by one of the authors, and specifically oriented towards developing associated sustainable development skills such as anticipation skills, strategic thinking skills, and risk awareness) that they found most useful, with one noting that it would be useful to have more modules such as that one which "challenged us to use our creativity and our practical reasoning" to address environmental issues which might "captivate more people and cause even more people to embrace this cause." One student found one of their management courses ("Crisis Management") to have been most valuable. Finally, another student made the important observation, key for those thinking about the relative value of "bolt on" versus "embedded" approaches to sustainable development education (McKeown 2002), that it was no "specific exercise that contributed the most" but rather noted that they had become "more responsible today because we continually returned to these matters."

Question 3

Students were more measured when affirming the applicability of the lessons learned to their professional lives, and their reservations hint at the significant barriers standing in the way of transitioning industry towards sustainability. A significant number of students (n = 9) strongly agreed that it was easy to translate what they learned into practice, with one student noting that at his company "they encourage us to propose new ideas." Nevertheless, many respondents (n = 21) found that it was possible, but difficult, to apply what they learned in their professional lives, with several respondents citing the gap between "theory and practice." Many students emphasized that companies "just don't seem to care" about environmental issues. Approximately half of these students (n = 10) explained that the difficulty of applying what they learned related to their current position within their companies. As one student explained, "until you have the power, it is hard to change things." Another agreed, explaining, "maybe one day with more responsibility I will be able to change that [the way his company handles waste]." One student pointed out that there was a strong social pressure to conform to the generally anti-ecological "uses and customs" at their company. Nevertheless, many students expressed an optimism that if we "continue to work on this subject in a self-aware way," we will eventually bring about an ecological transition by incorporating "environmentally friendly" changes into industry. Unfortunately, it is also true that a certain number of students (n = 6) expressed deep pessimism regarding the greening of industry. As two students noted, "the only objective of most companies remains profit."

Conclusions

Education cannot do everything, but we believe that our Circular Economy module is an effective tool for encouraging engineering students to become aware of both the need for more ecological and impact-sensitive forms of production as well as how industries are rising to this challenge. While we think that the role we as teachers play in this project is important, we believe that the majority of the learning actually happens as a result of interactions between students and their places of work as well as interactions between members of each team. We the teachers primarily serve the function of opening students' eyes to the reality of their workplaces. We prompt them to be inquisitive about aspects of their company that they might otherwise not have noticed, and we put them in a position in which students discuss the ecology of their workplaces with one another, setting them up, among other things, for the realization that there are connections between working in institutions that care about their ecological impacts, as well as those who work there. Over the course of this year-long project, students experience a whole range of what Albrecht (2019) calls "Earth emotions," a sense of curiosity regarding their company's practices, sometimes a sense of shame or frustration regarding industrial unsustainability, also at times hope and even pride regarding how their company, or those of their peers, are striving to make a difference for our collective future.

One widespread approach to thinking about engineering education for sustainability has been to focus on the development of key sustainability competences.[2] Unquestionably the attraction of this competence-based approach has to do with assessment: when one isolates a competence, one can test for it to determine the efficacity of a training program. It seems to us that the real measure of any education program devoted to ecology is never to be found in short-term assessments. We can—and do—test whether our students have understood the notion of the Circular Economy, a notion that itself implies that they have mastered the systems thinking competence that many studies (Mukhtar et al. 2019; Quelhas et al. 2019; Beagon et al. 2020) found to be vitally important for training engineers for sustainability. This will not tell us whether our students will have the resolve to take this knowledge and apply it consistently, over the long term and in the real world. It is only this care and resolve that will allow current industrial practices—which in many cases remain trapped in a paradigm where profit is the only motive force—to transition towards a more sustainable and ecological future. We will be the first to admit that we ourselves do not know whether our students will remain committed to bringing about changes twenty years from now, or whether they will become leaders in this transition. We do know, and this from their own testimony, that going out into the wilds of the industrial world and learning to attend to the different impacts of industry did, in the short term, prompt them both to learn about the ecological impact of engineering and, more importantly, to care about these impacts.

Notes

1 Please note that some of the responses quoted below were initially written in French and have been translated by the authors.

2 For example, two of the investigators are also participants in A-STEP 2030, an EU-Funded project aimed at determining the key competences needed to transition engineering education towards sustainability.

References

Albrecht, G., 2019. *Earth emotions: New words for a new world*. Ithaca, NY: Cornell University Press.

Arnsperger, C. and Bourg, D., 2017*Ecologie intégrale*. Paris: Presses Universitaires de France.

Beagon, U., *et al.*, 2020. *Engineering skills requirements for sustainable development and achieving the SDGs: Outcomes of focus groups held in Ireland, France, Denmark and Finland as part of A-STEP 2030 project*. Project report. Retrieved from www.astep2030.eu/en/project-reports.

Bellacasa, M., 2017. *Matters of care*. Minneapolis, MN: University of Minnesota Press.

Deudney, D., 2020. *Dark skies: Space expansionism, planetary geopolitics, and the ends of humanity*. Oxford: Oxford University Press.

Gilligan, C., 2016. *In a different voice*. Cambridge, MA: Harvard University Press.

Graham, R., 2018. *Global state of engineering education*. Cambridge, MA: MIT Press.

Held, V., 2005. *The ethics of care: Personal, political, and global*. New York: Oxford University Press.

Leopold, A., 1966. *A sand county almanac*. New York: Oxford University Press.

Marx, L., 2000. *The machine in the garden*. New York: Oxford University Press.

McKeown, R., 2002. *Education for sustainable development toolkit*. Paris, France: Waste Management Research and Education Institution.

Mukhtar, N., *et al.*, 2019. Environmental sustainability competency framework for polytechnics engineering programmes. *IEEE Access*, 7, 125,991–126,004.

Noddings, N., 2013. *Caring: A relational approach to ethics and moral education*. Berkeley, CA: University of California Press.

Quelhas, O., *et al.*, 2019. Engineering education and the development of competencies for sustainability. *International Journal of Sustainability in Higher Education*, 20 (4), 614–629.

Raworth, K., 2017. *Donut economics: Seven ways to think like an economist in the 21st century*. New York: Chelsea Green.

Shahidul, M., 2020. Engineering education for achieving sustainable development goals by 2030. *Science International*, 32 (4), 403–410.

Vogel, S., 2015. *Thinking like a mall: Environmental philosophy after the end of nature*. Cambridge, MA: MIT Press.

Wiek, A., Withycombe, L., and Redman, C., 2011. Key competencies in sustainability: a reference framework for academic program development. *Sustainability Science*, 6, 203–218.

12

EMBODIMENT AND MORE-THAN-HUMAN TOPOGRAPHIES

A Praxis Tool for Reconfiguring Sense of Place in the Anthropocene in Online and Limited-Residency Higher Education

Mary A. Jackson

Prescott College, a small liberal arts college in Arizona known for experiential learning in undergraduate courses, is at the forefront of limited-residency and online interdisciplinary graduate programs (Master of Arts and PhD). The newly formed Master of Arts in Outdoor Education Leadership (OEL)[1] is one such program at the college. Instructing graduate students in the field of outdoor education, one would assume, involves field-based and out of the classroom learning; however, these graduate courses are taught entirely online. Integrating experiential learning into online learning management systems moves students beyond the theoretical and builds recognition of human/more-than-human relationships and their own embodiment with place. The emerging theories of new materialism provide a methodological grounding for such encounters as students themselves venture into the outdoors in their own research and work to connect with place more intensely and intrinsically.

In this chapter, I present a creative praxis project applied in the core OEL course "Earth in Mind: ecological literacies for outdoor educators." This praxis project decenters human perspectives of place, integrates embodied awareness into experiential learning in more-than-human nature,[2] and reconfigures dominant anthropocentric narratives of nature and place in outdoor education. This signature assignment encourages students to examine their relationships with nature. The process, or what I call a *praxis tool*, is based on my field research in the Himalaya and the theoretical frameworks of new materialism. The assignment guidelines are introduced at the start of the course with a narrative and multimedia example, presented in part below and in the supplemental materials.

This praxis project and tool are not intended to be prescriptive but to engage in a contextual embodiment and experience of place. The nature of this tool allows it to be used in various disciplines and is based on five points of praxis on which the course is built: (1) embodied observation and awarenesS; (2) topographic storytelling; (3)

DOI: 10.4324/9781003221807-12

reciprocal relationships; (4) multispecies encounters; and (5) situated awareness in place and the Anthropocene. Instructors can use this praxis tool to empower students to examine their experiences as they relate to sustainability and to make positive actions to reconfigure enactments of the Anthropocene.[3]

"Earth in Mind" is an eight-week, three-credit course designed to foster the development of advanced ecological literacies, reflecting analysis and appreciation of the ways humans relate to their environments (ecological, spiritual, and cultural) and the relevance of these ideas to outdoor education. The course description explains that will challenge students to critically examine the broader social and ecological implications of outdoor education practices in various areas. Additionally, students examine their ontology and positionality as it relates to more-than-human nature and investigate new approaches to engaging in such environments. As such, students gain valuable insights to be more thoughtful and responsible in their roles as outdoor educators in the Anthropocene. Vital to reconfiguring dominant narratives and subversive practices is the inclusion of decolonizing methodologies and traditional ecological knowledge (TEK). Essential to the course is introducing theoretical concepts that challenge the dominant Cartesian assumptions of nature and culture as students begin to shift their perceptions of place.

The praxis project and tool are intended to guide students to break from the script, outcomes, or even the itinerary. The dominant goal-based narratives of the outdoors are inherently anthropocentric. The aim is to make linear goals secondary (or not at all) to reciprocity, reverence, and respect for the multitudinous relationships humans experience with more-than-human nature when embodied with a place, such as mountains. Perhaps the goal of getting to the top still occurs but in a different journey and process than anticipated, one in which contextual and locally relevant details matter.

Community and Praxis

A challenge in authoring "Earth in Mind" was to develop a pedagogical process that exposes students to the discourse and media of the Anthropocene and immerses them in the more-than-human environments of their own contextual experiences and homes away from their computers. These students live across North America and, occasionally, Europe and Asia. The flexibility of online learning provides geographic diversity as each student brings their subjective and local experiences to the learning community.

In the Prescott College graduate programs, instructors integrate specific project-based learning, or community based experiential learning (CBEL), into courses (Ramsey et al., 2020). CBEL projects ground hands-on, experiential learning—or learning by doing—within the students' communities. By connecting students to their communities, they can combine theoretical concepts and content with the practical aspects of learning. These experiences reinforce concepts and can offer a sense of engagement and depth to online courses (Greeson and Currey 2021). As such, the praxis project expands learning outcomes and emphasizes relationships and connection through sense of place within the conceptual frameworks of the course. Project locales have ranged from urban environments like Boston to the shores of Charleston, South Carolina, the Antero Reservoir in Colorado, to Jeju Island, South Korea, New Mexico deserts, and Southern California coastal mountains.

The CBEL for "Earth in Mind" is the praxis project with the goal to explore one's subjective experience with place. As explained to the students in week one of this course, students will do the following:

> complete a place-based creative project integrating their community and local more-than-human environments within a broad context of outdoor education, leadership, or outdoor/nature-based experiences. This will include a creative presentation and written narrative of one's relationship and experience with place through the lens of ecological literacies.

The students incubate their project relevant to the course content areas. They include both a written narrative vignette and a multimedia presentation, which may range from paintings and drawings to videos, slideshows, and spoken word. The guidelines allow for creative freedom in presenting projects. Students follow conceptual steps, or the points of the praxis tool that encourage them to shift their ways of thinking about place and how they are documenting it with both language and media. As such, each project is unique. I instruct students:

> Immerse yourself in more-than-human nature/environments in what you consider your community. This could be your town or backyard, or perhaps a mountain range or trail that you spend time in. It could be your local crag,[4] river, or city park. Develop a narrative of place and your experience based on the points of praxis (praxis tool). Through methods of ethnography and participant observation, I encourage you to document your personal experience/relationship with place.

They further examine and apply theoretical concepts and reflect on their project experiences through weekly required readings and forum discussions.[5]

Theoretical Concepts

Woven within the course is a significant amount of theory to frame the praxis tool and support an understanding of the reconfiguration of dominant narratives. I introduce new materialism and posthumanism but do not overload students with only theory. I may have them start by sitting outside and observing what they experience with their senses, followed by reading a chapter on Cartesian dualism. Additionally, I include summaries of theory as presented in this section. While theory may be a challenge for some students, I elucidate its practicality by connecting their forum discussion posts to field-based examples and/or their praxis project. The goal is to connect the theory back to how they understand place and apply that to their work as outdoor educators—or rather, to their locale and context. Theory is essential to the project as students unlearn what they know about the separation of nature and culture and unpack it in their subjective understandings about place, more-than-human-nature, and agency.

Posthumanism, New Materialism, and Agency

The dominant positivist, Western perspective centers human. They are the subjects that may classify and use (i.e., objectify) the world around them according to their supposed needs. In the postmodernist perspective, the world and meaning are constructed through language. Language is considered to provide the essence of what *is*, even though meanings may shift. In both situations, the rest of the world, the so-called non-human or nature, is understood in a binary definition as separate: man and nature, culture and nature, society and nature. The world in this shared paradigm is explained through a reflexivity that mirrors what *man* (i.e., dominant Western, white man) sees and interprets. Contrarily, posthumanism considers that humans and all phenomena of life are enmeshed and do not become what is real or essential separately but rather are entangled. As Barad (2007, p. 136) said, "refusing the anthropocentrism of humanism, posthumanism marks the practice of accounting for the boundary-making practices by which the human and its others are differentially delineated and defined." Posthumanism and new materialism counter the dominant Western perspective through which neither language nor matter are privileged and in which meaning is emergent. This idea directly addresses the dualistic separation of human and nature—one that is vital to think about and address the Anthropocene and, as such, is integrated into this course. Posthumanism asks not what phenomena in a static, singular definition are but rather *how* are phenomena (such as place) in a dynamic and subjective account of how the world becomes.

Drawing on traditions of twentieth-century critical theory, new materialism dissolves the mind-body dualisms prominent in traditional Western thought (i.e., continental philosophy). This process allows for a rethinking of human enactments within nature and the discourse and boundaries that seek to separate nature and culture. Shifting perspectives and conceptualization requires understanding how humans are entangled in the materiality of nature and culture. As Coole and Frost (2010, p. 1) explain, "our existence depends from one moment to the next on myriad micro-organisms and diverse higher species … as well as on the socio-economic structures that produce and reproduce the conditions of our everyday lives." This perspective decenters humans and requires us to understand that we are in a multispecies community, inseparable from not only other life such as bacteria to pets, but also from traditionally non-living matter of the world such as mountains, dirt, and in the material structures of power and politics. New materialism has connections to feminist theories, environmentalism, and ecofeminism, in which the dualities of nature/culture, animal/human, and mind/body are disrupted. These dualisms contribute to the conceptual disconnect of the Anthropocene and hinder our relationship to our embodiment with nature. Humans do not interact with the environment, people, thoughts, and things but rather *intra-act*[6] and through those actions embody the experiences through which meaning is made. These intra-actions shape the natural world just as the natural world we are part of shapes us. These concepts precisely examine the ideas of agency—not as a thing, but as a doing (Barad 2007)—of which more-than-humans enact. Barad's neologism, *agential realism,* provides one such framework for this examination. The key here is for students to see the intra-actions and agency as "a doing" within their contextual experiences. These concepts reorient one to confront the destructive aspects of living in the Anthropocene. For some students, it may be an emotional or confusing challenge as they unpack the topics and apply them to their context.

Agential Realism

Agential realism offers an account of the world as a whole rather than composed of separate natural and social realms, which is all at once a new epistemology, ontology, and ethic:

> The point is not simply to put the observer or knower back in the world (as if the world were a container and we needed merely to acknowledge our situatedness in it) but to understand and take account of the fact that we too are part of the world's differential becoming.
>
> *(Barad, 2007, p. 91)*

Thus, all phenomena have agency and influence each other in the dynamic states in which they are entangled. Matter does not seemingly exist separate from other matter, thoughts, memories, and actions but rather becomes what it is through the relationships within configurations of matter in space and time. While these concepts can be difficult to grasp for a first or second-semester graduate student, connecting them to their practical and real-world experiences helps to ground this theory in practice (praxis). For example, I present the following lines of inquiry in a forum discussion:

- Choose a place—wilderness, more-than-human nature, a location of outdoor programming you have worked. Consider this week's literature and the concepts of agency, Barad's agential realism, and performativity:
  - Based on Barad's definition of agency—in that it is not given but performed/enacted—in what ways does this occur in this place?
- How are the meaning/entanglements of this place different based on specific contextual configurations?
  - Identify not just the components of place, but the events—the emotions, actions, and things that occur that are not static and what is said and not said—a dynamic intra-action between human and more-than-human also looking at the words, the imagery among these things.

These questions may bridge a conversation in which a student perceives that the rock they have climbed dozens of times is more than inert, but rather that rock and their presence on it indicate dynamic social, political, and geologic forces they may never have thought of before. This may lead to them understanding the rock is on traditional Indigenous land and that it was once under an ocean or a volcano. Thinking differently about place can begin to shift their ideas of power and dominance or meaning and change.

Agential realism considers matter's dynamism and shifts focus from the ideas of reflexive (i.e., do they mirror nature) practices to the actions and relationships that make meaning and matter: the performativity[7] of matter (Barad 2007). Meaning is not an identity; meaning is material, and matter is not what exists separately from meaning. Meaning does not exist outside of specific intra-actions but only is made by contingent separations within phenomena. All matter (e.g., a rock one climbs) has agency and meaning entangled within all human thoughts, movements, history, and futures. Agential realism allows for a vibrant rethinking of the world, specifically in the lens of Western and continental philosophy that contends with established concepts of duality. Such a practice counters Cartesian dualism through an analysis that recognizes not a split of mind/body/spirit but the intermingling of such things. In this context, I encourage students in "Earth in Mind" to examine the existence of being (in place) as a dynamic configuration of space, time, and meaning. This practice counters long-held assumptions, particularly in outdoor education, of the meaning of activities, nature, environment, and so on. A mountain may no longer be a vehicle for human conquering and accomplishment but rather a reciprocal relationship wherein the goals of an itinerary are secondary to examining that experiential relationship. The praxis project integrates these theoretical concepts with their own contextual (and local) experience of place as they put into practice reconfiguring the discourse of nature in a way that engages with the agency of themselves, of place, and the meanings they enact within outdoor education.

Praxis Project

The praxis project demonstrates entangled human and more-than-human relationships that are vital to the meanings and experiences of place. I developed the praxis tool through my research in the Himalaya using methods of a "walking" ethnography of adventure trekking tourism in and around the trails of Everest Base Camp Trek, Khumbu, Nepal. A walking ethnography is simply a method of understanding the nature and culture of a place by "being" in place, which may include being stationary and observing or moving about. My research in Nepal examined research and narratives that decenter human and goal-based experiences of the outdoors and focus on the dynamic and entangled relationship among human and more-than-human. This entanglement of place blurred the lines of human history, colonialism, geology, and glacial melt, revealing the inseparability of such phenomena. With the following narrative vignette, which frames the start of the course and project, and further discussion and readings in class, students learn to shift their observation of place, emphasizing an embodiment[8] with place. The following narrative excerpt (Jackson 2017), along with multimedia videos, provides context and examples for their projects. I use this vignette as an example as it frames many of the challenges of outdoor education and a complicated history of colonial and anthropogenic approaches to adventure. Everest Base Camp, for example, is a type of ground zero for the outdoor industry, and such an example is relatable to students in our program. This example demonstrates agency, intra-action, and performativity.

Dengboche, Nepal (14,469 Feet, 4410 Meters)

Woven trails guide us through the lowest points of the valley. The trail emerges higher, surrounded by seemingly taller peaks than seen before. The dusty paths are carved by feet of locals, travelers, nomads, traders, tourists, and the yaks and horses that bring so many curious minds and bodies (mindbodies) to this place. It is carved from those desperate for money, those out to prove something to themselves or others, those with an innate thirst for adventure or excitement, those seeking work, those going to work, those who want to know more about why people are here, why this place. And myself, questioning what these Himalayan mountains, and specifically Mount Everest, mean to humans and to understand these mountains with agency and life in their own right.

These paths, however, are not solely made by humans. These paths exist because of the movement and reaction of dirt and soil and rocks, and from those iterations of human movements. These paths exist from economic and cultural needs and wants of humans; from the imperialist symbolism of early British expeditions, the romantic hyperbolic and post-colonial expeditions and the inward drives to climb, the commercial commodification of name brand expeditions, to the hypertouristic movements that lead so many to the base of (and up) Mount Everest. This path is entangled with weather systems that shape and erode the valley, from the climate patterns of the subcontinent to the geophysical history and the glacial movements that shape and twist the yul lha (peaks) and beyul (valleys) of Khumbu.

The entire landmass of India is pushing up against these mountains; Everest itself is still growing. These mountains are young. As a child, while walking in the Appalachia of Virginia, I was told that they (the Appalachian Mountains) were the "grandfather mountains" and were once the size of the Himalaya. These mountains we are in now, the Hindu-Kush Himalaya, are the youngest in the world. Moving like the growth plates in my own adolescent son; so slow from a human perspective of time, but in a planetary and cosmic sense a blink of the human eye.

This path guides us today. The throngs of people move slower than my liking, I attempt my usual leapfrogging. I slow down at some points to speak with English speaking tourists and realize that no one wants to talk much while hiking at this altitude, aside from usual pleasantries of hello, where are

> you from, where are you going. But at this point in Khumbu, most people going up are going to EBC. Some however, will stop at Lobuche Base Camp (or move there after staying at a Lobuche guesthouse) for the two-day ascent of Lobuche (an acclimatization for climbers going up higher peaks or for a goal in and of itself). At a juncture in the trail, in Nepali I ask a Sherpani moving yaks along for directions to Lobuche.
>
> *(Jackson 2017, p. 242)*

This vignette aims to take theory and implement it into practice by demonstrating how one can reconfigure a dominant narrative of place and integrate dynamic meaning and performativity. Place is not static; humans are not centered. Specifically, in outdoor education, place is often commodified as a tool for human experience and achievements (e.g., reaching the top of Mount Everest). Shifting ways of thinking and integrating the agency of more-than-human nature while dissolving dualisms of nature and culture is a way of decentering humans and recognizing the entanglements of place. Students are encouraged to use this shift of perception in how they approach teaching outdoor education and expand from there to other sustainability practices in their professional, academic, and personal lives. Theoretical concepts may be empty without direct application.

FIGURE 12.1 Freeze–thaw cycle noted in the early morning below Cho La (pass), Khumbu, Nepal. April 2016.
Source: Mary Jackson.

FIGURE 12.2 Dirt, rocks, and peaks: an entanglement of place and reciprocal relationships, Khumbu, Nepal. April 2016.
Source: Mary Jackson.

Methods of the Project

I present an outline of the praxis project and tool in the first week of the course, along with the above narrative vignette and multimedia examples. Students are instructed to use walking ethnography and participant observation methods by documenting and analyzing their experience of place. These methods build into the theoretical understanding of the course and praxis tool. The following sections present how I introduce these methods to students.

What Is Walking Ethnography?

Walking ethnography involves the process of entering the field, "being in it," and leaving it. Yi'En (2013) explained that this involves the inclusion of both field sites and those things that are usually considered in the background, the distractions that can sometimes contribute to both the poetics and politics of field research (Vergunst and Ingold 2016). These hidden stories can illuminate the intricacies and contexts of place and experience. Yi'En said:

> Through viewing the act of "walking" as a mobile and embodied practice, I argue that "walking" is inherently a rhythmic experience and potentially offers insights to the multiple splices of time-space narratives. However, walking ethnography is not only constituted through "walking" per se *but also through the sensorial aspects of our bodies such as seeing, hearing, and feeling* [emphasis added].
>
> *(Yi'En 2013, p. 2)*

The process of taking notes involves moving through land, sitting in place, and being present in a place. This process is a rhythmical bringing of bodies into conversation with the environments humans move through. It is important to note the methods and intentions of walking ethnography do not require the ableism necessary to walk or hike. While that is what bodies may do in such places, the intent is how one's body embodies, reacts, and responds in a relationship with place. In week two of the course, as these concepts are beginning to take shape, I instruct students to "go outdoors into more-than-human nature. This could be a patch of grass near your home. Your backyard, the ski slopes, a city park, national forest, and so on. Expand your observational senses and reflect on the descriptive language you use." In the forum discussion, I ask, "How can you decenter the human perspective of place and describe the more-than-human nature surrounding you, a part of you (remember, even human bodies are nearly 40% not human!)?" This, and other such assignments, creates a foundation for how they go into the field for the praxis project and provides a lively forum discussion among students with subjective experiences. In addition, they have already read chapters 1 and 2 of *Subversive Spiritualities* (Apffel-Marglin 2011) that presents and grounds theory with examples.

What Is Participant Observation?

Participant observation is a term given to methods of immersing the researcher into a field site and gathering data through both observing and being a part of the study group. It is linked to various qualitative research studies in anthropology, social sciences, sociology, and education. Data is collected through field notes, surveys, and other forms of documented experiences (DeWalt and DeWalt 2010). It means immersing yourself in an environment or place, noting what you see and observe, and documenting your interactions with the place. In this context, it may mean casually interviewing others, journaling your experiences, or sitting and watching the world around you. Students begin to do this with the forum discussions and then on their own while making their project; they might wander the sandy back alleys of Pacific Beach, San Diego, or ski the slopes of Copper Mountain,

Colorado. Regardless of where, they begin to observe the intimacies of place, their reactions to it, and responses from the land underfoot.

Introducing participant observation to students encourages them to understand they are not a passive observer but a dynamic actor in the configuration of place. Drawing on Barad's (2007) agential realism, one is never separate from that which they are observing, and one's presence impacts the outcomes of the observation. This method encourages them to write on their subjective experiences, emotions, and feelings entangled within the environment and other beings of place. They use many documentary tools including video, voice recordings, handwritten notes, interviews with other humans, and to seeking to understand or "interview" something such as a mountain, thus encouraging an embodied participant observation.

The Praxis Tool

The points of praxis presented in this section further outline how students *do* the project. I present these steps along with the above narrative vignette and methods. Important to note is that this process is not necessarily linear but, for the sake of clarity, listed from 1 to 5. The praxis tool guides the process of embodied participant observation and writing the narrative that aims to decenter the human. These bring to the foreground a set of urgencies and reminders for being in place in the Anthropocene and recognizing a more hopeful future. Utilizing the praxis tool requires an instinctual awareness and an intimacy with place. Fundamentally, it is not used to get to some point or destination or even conclusion. It is about the process and a framework for assessing and analyzing the experience of a contextual place in space and time. Key to understanding how to do this are the weekly forum assignments and the examples presented at the start of the course while immersing themselves in place. If students skip readings and forum discussions, they may lack a depth of conceptual understanding.

These points of praxis guide students to engage in the creative and field-based aspects of their projects. Being acutely attuned to the specific contexts and agency (entanglements) of place allows the praxis to unfold and be effectual. Instructors can include these steps individually in weekly forum discussions to examine their process and topics, or students can use this as an overall reference as they complete their projects.

Embodied Observation and Awareness

This point of praxis encourages embodied observation of intra-action and awareness of reactions and responses to place. This concept encourages walking ethnography through embodied observation in which humans are not just participating but intra-acting with entire bodies, senses, and minds in a place. Embodied observation takes cues and notes from what is happening to, with, and around the body. It attunes to the inanimate agencies of the more-than-human in a specific context of place. Embodied observation is a reaction to food and water, a reaction and response to altitude, temperature, breath, wind, and so on. In the mountains, specifically, it attunes to feet on or off trails, shifting and moving and responding to the land through an embodied communication. It is an opening of human awareness of being in place. Here a student opens their narrative with this awareness:

> The Superstition Wilderness is vast and rich with biodiversity that penetrates every sense of being upon arrival and through its landscape. From the first steps of Siphon Draw to Flatiron Peak, the hiker is placed in a mindset of humbled awe bearing witness to the great Superstition peaks. The trek begins in steady incline, only to become more and more vertical deeper into the mountain themselves. Barad's agential realism becomes practice as the hiker finds themselves in denser rock, climbing in and out of crevices and reading meanings through their intra-actions with the desert.

This point should not come as a surprising awareness to practiced climbers or mountain travelers. This awareness is vital to keeping one alive and safe in the mountains. In praxis, it is noted and acknowledged and cultivated to a more intimate understanding of place. Embodied awareness draws on this recognition, which can highlight agency of the more-than-human. For students to engage in this aspect of the tool, an understanding of individual and personal embodiment with place is necessary, which they build upon in weekly assignments and readings.

Topographic Storytelling

This point necessitates students to ask many questions of place and experience. Responding to land and discussing the embodiment of place—not what it is, but *how* it is in dynamic discourse (not relegated to only words)—is a key of topographic storytelling. One must ask questions to learn, understand, and then tell stories of place: What are the entanglements of meaning for place through various and multiple perspectives? How are the meaning and multiple identities of place told through Indigenous cosmologies, stories, folklore, oral histories, traditional ecological knowledge, religion, and sacred landscapes? What does the scientific method teach one of these places (through disseminated science literature)? What do other disciplines share of this place, such as anthropology and social sciences? How do the arts respond to this place and make meaning through visual arts, audio, sculpture, music, and so on? What stories are told of this place and how? What do these stories express about this place, and how do you learn more about it that way? How are these stories anthropogenic, colonizing, or destructive?

Topographic stories are not *what* humans do but what humans have to do *with* place. Through an ethical and philosophical awareness and a remembrance of Indigenous and traditional ecological and cosmological knowledge, the destruction caused by the Anthropocene may be limited. In this technological adolescence of Westernized humans, where humans have expanded and explored across the planet, to our moon, Mars, and through the solar system, it is vital to understand the interconnectivity of life. This awareness can be developed by recognizing TEK and utilizing this knowledge in contemporary frameworks. Such a framework may reorient various levels of human action, enabling a reciprocal relationship between humans and more-than-human in the Anthropocene. This line of thinking is encouraged through each week of the course and included in forum discussions. Students examine the multilayered and dynamic history of place, like the example narrative of Khumbu in Nepal or this excerpt of a student's narrative on Jeju Island, South Korea:

> I wanted to move through the place and be responsive with reciprocity in Hallasan ... looking at intersectionality at the history, ecology, culture, and people to become immersed in the place of Hallasan. This ... led to uncovered realizations of imperialism, Western perspective, Shamanism meanings, the Anthropocene, agency of power, and climate.

Students must continually ask questions of the place of their praxis project.

> Jeju people have had to experience a multitude of invasions and resistance efforts ... Jeju people have spent the last 5,000 years struggling to survive in the harsh environment and subjugation by others. From the Indigenous Tamna with Shamanistic rituals and beliefs to the Koreans and foreigners living on Jeju today, Hallsan is shaped by many of their stories and meanings of Hallasan.

This narrative invited one to then view a multimedia video and concluded:

> As we learned, the destinations were second to building a relationship with Hallasan. I hope that you can hear, see, and feel the ancient forest speak as the Koreans say, "You humans are part of nature too, and so you should adopt the ways of nature. It tells us to be humble."

Reciprocal Relationships

Reciprocal relationships between humans and more-than-human involve understanding informed consent of place. It means going into place with an awareness of the nuances, the agency and scale of place, skill and movement through place, and the various responses and reactions from the human body to more-than-human. Knowing place means learning about it, the stories, the features, the embodied experiences. Allowing place to respond and teach us to move within enables respect and reciprocation. This includes learning the nuances of ecology and geology, the weather systems, where the sun rises and sets, where the water runs or does not, how it changes in moments of the day or seasons. This does not come as a surprise or challenge to many who adventure into places, as being attuned to objective hazards and topography is essential to risk management in the outdoors.

As such, OEL students may be primed to expand this awareness to a deeper embodiment and awareness as this student introduced in the start of their narrative:

> When you're hiking through the redwood forests in Santa Cruz, every sound is muffled, with the exception of your footsteps, your breaths, and the songs of birds that are in a close proximity. There are plenty of other people playing in the forest as well, including runners, hikers, and mountain bikers, but the sounds of their footsteps and bike tires quickly fade away after they pass, once again leaving you in solitude. However, when you stop to really listen carefully, you can hear the rustle of a bird's wings high above you in the air, and even the sound of an eagle turning its head while it surveys the ground for rodents below. In addition to the sounds, you can see plenty of evidence of geologic time which helps put our place in the universe and existence into perspective.

Learning the intimacies of place allows for the humbling knowledge of human's role in place to emerge. Some may notice, over multiple visits to the same place, that small changes occur, such as ice slowly freezing over on a lake (and perhaps later in the season in recent years), or how the shape of a trail changes with the light of the day, and how snow shifts the experiences of familiar places changing how they move about on the land. Others may notice how their breath and movement change with the topography and altitude or how their knees ache on a descent. Sometimes, these intimate aspects allow one to begin to understand their relationship with place more intimately.

Multispecies Encounters

Multispecies encounters mean being cognizant of and open to the relationships humans have with Earth's many species and how those species are integral to experiences and life. Like with topography, this means being aware of the reactions, responses, and kinship of these species to the entangled experience of place. Multispecies encounters are about awareness, subversion, and action in an entangled world. What other species teach humans through encounters and relationships allows humans to understand themselves as dynamic actors in the diverse configurations of nature and culture. This analysis allows humans to be more aware of reactions to other species and examine and change behavior. It can help pose questions that challenge assumptions of what it means to be human. Multispecies encounters can lead to an understanding of the agency of more-than-human and unnoticed landscapes. In this essence, such planetary infrastructures, like mountains, are far more than static background structures of the human story.

To develop this awareness, students examine human and more-than-human relationships. They write a short "multispecies ethnography" using the following prompt:

> Inspired by this week's readings, write a short multispecies ethnography utilizing this theoretical framework. This could be between you and your dog,

you and the soil of your garden, you and the rhizomatic aspen groves, you and the hard rock of a climb, and so on. You are encouraged to also include other media in your post, such as drawings, photos, etc.

This assignment practices reconfiguring their conceptual awareness of being human and supports putting theory into practice. Students often examine in more depth their relationships with a pet, such as their dog or cat, and sometimes expand this to their entire project in which their dog may experience place inseparable from themselves.

Awareness of Place and the Anthropocene

Sustainability and the Anthropocene can be overwhelming. In discussing sustainability, climate change, and the politics, policies, and activism that go along with that, it is easy to succumb to the big picture, to the longitudinal change and observations, to a concept that "more is better." Situated awareness is what allows us to understand the magnitude of the Anthropocene, but it also can do the opposite—shut us down, overwhelm, consume, and saturate without getting to a core understanding of something that can help change or cope with that state of the planet. Taking a slice of an experience and a place, one that is both a dynamic moment in time and no longer present, can help develop a situated awareness of place and the conditions therein. I encourage students to examine how this can be achieved in outdoor experiences.

Throughout each of the assignments in this course, and while completing the praxis project, I ask students to slow down. In this situated awareness of place, one may stop, slow down, simplify, and notice the inanimate intimacies of place. Through embodied participant observation, one may talk, look, listen, feel, and hear. One may notice how a river braids down a valley from a glacier, notice the sparkling dust and light, and not seek to move up or down, but to be in place without an end-goal. To encourage this, one may limit the goals of going up and the destination as important. Destinations are still a point but are secondary to agency and presence of place. This tool is not about going to mountains or more-than-human nature for revelation, to seek a deeper understanding of self; the agency of mountains is not a tool for human ego-expansion but rather to understand the sublime and often precarious relationships among beings.

Pedagogical Outcomes and Impact

This praxis tool and, to a greater extent, the curriculum of "Earth in Mind," guides a process of observation that aims to bring more-than-human nature into the center, play, and prominence. Through this process, a reciprocity develops in which human among more-than-human becomes embedded and inseparable. Shifting an anthropocentric gaze that privileges and sets matter apart as isolated and constrained by boundaries determined by humans demonstrates the vibrant agency of more-than-humans such as forgotten landscapes, rocks, dirt, and glaciers, to the tiny bugs and creatures inseparable from human experience. In this perspective, nature is not a passive object upon which humans descend but rather an entangled subjectivity. This acknowledgment allows for a rethinking of human enactments of the Anthropocene and complicit behaviors of this epoch, reframing approaches to relationships with nature both in and out of the classroom.

Acknowledgment

This chapter was written on traditional lands and territories of the Núu-agha-tʉvʉ-pʉ̱ (Ute), Tsésthoʼe (Cheyenne), and hinonoʼeinoʼ biitoʼowuʼ (Arapaho) in contemporary Golden, Colorado, and in Solukhumbu, Nepal, the traditional and continued home of the Sherpa. Prescott College is on traditional lands and territories of the Yavapaiv Apache, and Hohokam.

Notes

1 Formerly the Master of Arts in Adventure Education.
2 I use more-than-human to refer to conventionally living and non-living matter that is not human.
3 Course syllabus, discussion prompts, and literature included on the companion website.
4 Colloquialism for rock climbing cliff
5 See the companion website.
6 The relationship, reaction, and responses of phenomena. Through specific intra-actions, boundaries and properties of agents become determinate. (Barad, 2007)
7 A departure from linguistic definition of performativity, it has to do with the enactments of meaning and identity (Barad, 2007).
8 Embodiment is the transcorporeal experience of place, that time-space where human corporeality, in all its material fleshiness, is inseparable from "nature or environment" (Alaimo, 2008, p. 238).

References

Alaimo, S., and Hekman, S. J., 2008. *Material feminisms*. Bloomington, IN: Indiana University.

Apffel-Marglin, F., 2011. *Subversive spiritualities: How rituals enact the world*. New York: Oxford University Press.

Barad, K., 2007. *Meeting the universe halfway: Quantum physics and the entanglement of matter and meaning*. Durham, NC: Duke University Press.

Coole, D. H. and Frost, S., 2010. *New materialisms: Ontology, agency, and politics*. Durham, NC: Duke University Press.

DeWalt, K., and DeWalt, B. R., 2010. *Participant observation: A guide for fieldworkers*. Walnut Creek, CA: Rowman Altamira.

Greeson, K., and Currey, R., 2021. Centering justice in a sustainable food systems Master's program. Retrieved from www.frontiersin.org/articles/10.3389/fsufs.2021.751264/full (accessed November 17, 2021).

Jackson, M. A., 2017. Process and emergence: A topographic ethnography of the embodiment of place and adventure tourism in Khumbu, Nepal. PhD thesis, Prescott College.

Ramsey, S., *et al.*, 2020. 2020 vision: How a global pandemic and the Black Lives Matter movement focused our teaching. *European Journal of Educational Science*, 7 (4), 27–39. http://dx.doi.org/10.19044/ejes.v7no4a3.

Vergunst, J. L., and Ingold, T., eds., 2016. *Ways of walking: Ethnography and practice on foot.* Abingdon: Routledge.

Yi'En, C., 2013. Telling stories of the city: Walking ethnography, affective materialities, and mobile encounters. *Space and Culture*, 17 (3), 211–223. doi:10.1177/1206331213499468.

13

INHABITING SOUNDS

Soundscape Ecology in a First-Year Seminar

Damiano Benvegnù

In his 1982 book *Language and Death*, Italian philosopher Giorgio Agamben explores Aristotle's famous statement that while the other animals have only voices, "man alone of animals possesses speech" (Agamben 1982, p. 1253). In the epilogue of this work, Agamben describes the acoustic impression one might receive while walking through a forest:

> It happens when we walk in the forest and, suddenly, unheard, the variety of animal voices surprises us. Whistles, trills, clucks, beats coming from wood or broken metal, shrills, rattles, whispers: each animal has its sounds that originate immediately from it. Ultimately, the double call of cuckoo bird jeers at our human silence and exposes, unbearably, our being the only creature without voice in the infinite choir of animal voices. Then, we try to speak, to think.[1]

Here, in referring to such an immersive nonhuman soundscape, Agamben seems to flip the common understanding of nonhuman muteness on its head: the idea that the nonhuman world is fundamentally silent—nonlinguistic—is reversed to claim instead that our human *logos*, our peculiarity as creators of written discourses, emerges from a loss of voice. He maintains that human language is the grave of the animal voice, where the voice of the animal is both arrested and preserved (Agamben 1991, pp. 45–46). His description of the acoustic experience during a forest walk thus fits within a broader attempt to express the link between anthropogenesis and what Agamben calls an "*experimentum vocis*," through which we, as humans, "can radically question the role of language in the voice and try to assume being a speaker anew" (Agamben 2018, p. 23).

While I do not explore all the philosophical implications of Agamben's onto-linguistic analysis, I often use the description of the soundscape he provides at the end of *Language and Death* in my environmental humanities classes. I employ such

DOI: 10.4324/9781003221807-13

excerpts to introduce my students to a different kind of experiment, involving soundscape ecology and the importance of nonhuman sounds for rethinking environmental consciousness and ethics. Agamben's example highlights the alleged distance between human linguistic practices and nonhuman semiosis, while being at the same time a quite evocative written text. This latter aspect became crucial in a first-year seminar I taught during the 2019 Spring term at my institution, Dartmouth College, because these seminars are meant to be structured around intensive writing but within a specific discipline. In this chapter, I articulate the significance of introducing soundscape ecology practices in my first-year seminar, describing the logistical, pedagogical, and practical steps I take to lead my students to experience both the ecological importance of sounds in the physical environment and the difficulty of translating nonhuman sounds into a written text.

The Course

First-year seminars at Dartmouth stress the importance of written expression while providing first-year students an opportunity to perform research within a specific discipline. They are hosted by Dartmouth's Institute for Writing and Rhetoric in collaboration with an academic department. My seminar, "Environmental Italy: Narrative, Landscapes, Ecologies," was part of the curriculum of the Department of French and Italian, but it was also designed to serve as an introduction to the environmental humanities through a specific geo-cultural lens, i.e., Italy. In fact, "Environmental Italy" engaged students with what Italy can teach us about our Western relationship with the environment, using Italian ecological narratives as a comparison to reflect upon environmental knowledges and practices in the United States.

The course therefore had two intertwined pedagogical trajectories. On the one hand, we stressed how Italian landscapes have been historically recognized as repositories of stories, conveying narratives of environmental resistance and ecological liberation as well as embodying the historical continuity between human communities and specific territories. Students who took the class learned how Italian writers, scientists, artists, and philosophers imagined and represented real and fictional environments and how these representations reflect, critique, and animate the approach that Italian culture has had toward the physical environment and its ecology since the Middle Ages.

In addition, "Environmental Italy" had an experiential learning component designed to enhance its comparative goals. Thanks to a Humanities Lab Grant from Dartmouth's Leslie Center for the Humanities and the support of Dartmouth's Center for Social Impact, my class paired up with the local non-profit land conservation organization Upper Valley Land Trust (UVLT) for a "Social Impact Practicum," which is a project-based experiential learning opportunity through which Dartmouth students take the skills and content that they are learning in their courses and apply them to relevant projects for, and with, local community partners. In the "Environmental Italy" case, my familiarity with the UVLT and their work on land preservation allowed me to organize a series of field trips meant to encourage my students to shift from the textual and visual artifacts of the distant culture we analyzed in class to the physical reality of a specific nonhuman world nearby, as a way to recognize both the transition from the representational to the physical (and vice versa) and the potential eco-cultural differences between Italian approaches to the environment and those surrounding our own Dartmouth community. One of these field activities brought my students to the Lyme Hill Conservation Area, 238 acres of mostly forested, hilly landscape on New Hampshire's Route 10, some ten miles north of Dartmouth's hometown of Hanover. This conservation area is owned by the Upper Valley Land Trust and has been managed for a mix of recreation and wildlife habitat over the years. Most importantly for our field activity, outside of the forest and trails, there is a 3.6-acre meadow along Route 10, which also comprises a marsh area. The presence of these different environments within a relatively small property creates several "soft edges," or transitional areas between habitat types, that appeal to wildlife and are usually rich in biodiversity. The Lyme Hill Conversation Area is rich in ecotones, those regions of transition where two or more biological communities meet and integrate, often with the process of juxtaposition, intermingling, and fluid interactions.

Ecotones are crucial elements for the specific assignment to be performed at Lyme Hill, as we were determined to listen to, record, and then try to transcribe the soundscape of different parts of the natural area under examination. In fact, ecotones not only characterize the health of a biome and determine its biological and semiotic richness, but they have also been compared with cultural contact zones, those "social spaces where cultures meet, clash and grapple with each other" (Pratt 1991, p. 34; cf. Eppelsheimer et al. 2014). This overlap provided a deeper meaning to the assignment as our class was also meant to move beyond monoculturalism and monolingualism and establish instead a fruitful collaboration between transcultural consciousness—the intentional self-reflectivity that expands our repertoire of cultural behavior and helps us behave adaptively in a different cultural context—and the environmental humanities.

The Activity

Before

Dartmouth College functions on a quarter system, so my "Environmental Italy" course was only ten weeks long. Every week we explored a different ecological theme or concept, moving from an analysis of what common environmental terms (such as "nature," "animal," "landscape," etc.) have meant—and still mean—both conceptually and socio-historically for Italian culture as well as for our own American contemporary society, to arrive eventually at a renewed, multi-faceted, and comprehensive understanding of the entangled relationships between the human and the nonhuman world. Throughout the course, we also returned regularly to the etymology of the term "ecology" (from the Greek *oikos*, with the meaning of "house, dwelling place, habitation," and the Proto-Indo-European root **weik-*, "clan") to discuss the link between environmental, geographical, and socio-political awareness in a more-than-human world, between multispecies places, practices, and communities. This quasi-etymological methodology allowed us to track the historical trajectory of crucial concepts in the environmental humanities while maintaining a comparative and experiential approach. Each week we had three lessons lasting for 65 minutes.

In week 7, the overarching theme was "Sounds," and we began the week watching and analyzing a 2010 Italian film entitled, *Le Quattro Volte* (*The Four Times).* *The Four Times* can be described as an experimental movie as it depicts the concept of metempsychosis taking place in the remote mountain town of Caulonia in southern Italy. In fact, while the first of the four movements or phases that compose the film features as the protagonist an elderly shepherd who herds goats and then gets sick and dies, the other three instead follow the alleged transmigration of life from the human being into a goat, a tree, and charcoal, respectively. Crucially, there is virtually no dialogue in the film, and the cyclical rhythm of life as suggested by *The Four Times* is accompanied only by the sounds emitted by the bodies of the different beings engaging with their surrounding environment. Along with the film, students read a chapter of Elena Past's *Italian Ecocinema* devoted to *The Four Times* and a brief article by Brian Pijanowski et al. entitled, "Soundscape Ecology: The Science of Sound in the Landscape" (Past 2019; Pijanowski et al. 2011).

The Four Times is a perfect introduction to sound and sound ecology precisely because it invites viewers to switch their focus from the human voice and its content to the immersive aspects of more-than-human soundscapes. As sound recordists Benvenuti and Oliviero pointed out, in fact, "when you remove the dialogue, you discover an enormous variety of sound" (Past 2019, p. 123). Moreover, the lack of apparent human action in the movie attuned my students to attend to otherwise neglected aspects of the reality depicted in front of their eyes, or, in other words, to react less to visual stimuli and instead to pro-actively search for other semiotic clues. Finally, and equally importantly, the film is located within an agro-pastoral world, thereby linking the importance of sounds not just to the environment itself but also to the specific socio-cultural practices performed in that environment. Although we did not dive into Italy's complex multilingual landscape, I told my students that human communities like the one depicted in *The Four Times* would likely not speak the standard Italian of the literary tradition in their daily interactions, but their own local language, usually characterized by an exclusive oral transmission.

In our in-class discussion, we talked about the importance of sounds in the film as a whole, as well as selected specific passages, and tried to describe in detail what we were hearing and what kind of semiotic processes were carried by the sounds recorded. I then asked students to discuss in groups the importance of sounds in their life and to try to come up with an acoustic geography of our campus, starting with what sounds they heard on that very day before coming to class, their specific location, and meaning. Finally, we talked about how all these sounds are both present and often obscured by what ethnomusicologist Steven Feld has called "the visualism deeply rooted in the European concept of landscape": our tendency to reduce the multisensory character of our perceptual experiences into an almost exclusively visual conceptualization of place (Feld 1996, p. 94). This analysis allowed me to introduce students to concepts and practices that might otherwise have been a bit too abstract, such as soundscape ecology and ecosemiotics. Both the study of the acoustic relationships between living organisms, human and non-human, and their environment (i.e., soundscape ecology) and "the study of the semiotics of relationships between nature and culture" (i.e., ecosemiotics; Kull 1998, p. 350) engage with the entanglement of nonverbal communications among different beings and their importance for the whole ecosystem. Ultimately, they both suggest that all living matter exists in a common communicative sphere and that there is a relationship between the health of a biome and the auditory signals coming from the creatures living in that environment.

During

The day before our field trip my class went to Dartmouth's Jones Media Center to get accustomed to the technology. There, my students got an introduction to how to use a professional audio recorder. I then divided them into pairs and gave each pair an audio recorder (either a Zoom H5 Audio Recorder or an Olympus LS-100 Audio Recorder) to bring home and experiment with before our field trip. The Jones Media Center loaned out these recorders to my students for three full days.

The following day we were bussed to the eastern entrance of the Lyme Hill Conservation Area. There, the UVLT ecologists assigned each group with their specific environment to record: either the meadow, the marsh, the creek, or the forest. They also gave my students both an introduction to the ecological features of their assigned environment and some hints as to what sounds to pay attention to. I then reminded them of the assignment: to listen to and record the soundscape of different parts of the natural area under examination, without interfering with their own voices, if possible. Finally, I instructed my students to go to their locations and just listen for at least 5 minutes before starting their recording. The total time for the specific assignment was 30 minutes (5 minutes of silence plus 25 minutes of recording). The time constraint was contingent (students had other classes to attend after "Environmental Italy") but also pedagogically intentional, as I encouraged them to be proactive and analytical in their listening. In other words, I wanted my students to engage with the surrounding environment with both general curiosity and a sense of purpose, balancing the awe of listening to a soundscape never heard before in such high-fidelity with a limited amount of time to collect the acoustic material they needed for completing the activity. I believe that such balance is a good ecological practice, as it breaks down the division between pure enjoinment and scientific interest; irrational wandering and strategic planning; and randomness and ecological knowledge (Benvegnù 2019, pp. 196–197).

The main goal of this part of the activity is to accustom my students to both their own silence and the sounds of the nonhuman world. In other words, my aim for this first installment of the activity was to counter our cultural tendency to attribute a constitutive silence to the nonhuman world as well as to introduce students to a different, richer practice of listening, what Murray Schafer (1992) has called "sound education." Moreover, listening walks such as the one we did on Lyme Hill can have a substantial phenomenological impact: they allow students to experiment—through sounds—with their own movement (or stillness) and their own senses and thus create "the possibility … to interrogate their habitual responses to the world, to offer bodies the potentiality for recomposing their corporeal relations to each other, to their environment, and to the ways that we experience and create knowledge" (Springgay 2011, p. 640). For instance, when asked to simply walk through the forest and be mindful to the sounds surrounding them, some of my students appeared to be initially uncomfortable with performing such an activity in front of the whole class. However, after a few minutes, they transitioned toward a different state of attentiveness, often repositioning their body to hear sounds better on the ground level or touching the water to test any potential acoustic alteration. Ultimately, my purpose was to expose students physically to the richness of life as embodied and represented by the whole system of sounds present on Lyme Hill as well as in any given environment (Murray Schafer 1992). This kind of listening does not reduce us to mere receptors, nor does it cause us to lose the ability to reflect upon what we are doing. Rather, listening walks both invite "listeners to listen to their own listening" and provoke "them to examine everyday habits of attention that are normally taken for granted" (Gallagher et al. 2017, p. 1248). This combination of phenomenological presence and meta-reflectivity was meant to provide students with a practical understanding of agency and cognition that is at the same time deeply entangled with the nonhuman world and yet still meta-cognitive and, therefore, quite speculative.

On the bus back to campus, I inform my students about the second part of the assignment, which is both practical and theoretical. First, I ask each pair of students to listen together, if possible, to their own recording; to select a one-minute sample of the recorded soundscape to upload on Canvas (our class portal) and present to the whole class during our next lesson; and to write down jointly a brief text (350 words) describing through words their soundscape. Second, I invite them to reflect upon the phenomenology of their experience: how their bodily perception of the forest changed through listening to the soundscape; the ecological value of the specific soundscape bits they recorded; and how our literate culture can engage with as well as properly dwell within the acoustic semiosis occurring in the nonhuman world. I did not ask my students to write their reflections down, but I urged them to come to class prepared to discuss their reflections with their classmates.

After

On our last class for the week, each pair of students have a few minutes to play their recording sample and read aloud their own transcript of the soundscape. I ask them to read the transcriptions aloud because I want to underline that even human communication has a crucial oral quality which is attached to our own bodies, a characteristic that might be overlooked in literate societies such as ours, where writing is often—and quite paradoxically—perceived as a disembodied activity.[2] I then allow the whole class to comment briefly upon what they heard—what sounds they recognized, what they did not recognize, how efficacious the soundscape transcription was, etc.—and to ask questions, if they had any. Unsurprisingly, different approaches to the writing assignment become apparent, ranging from merely listing the sources of the sounds they heard and selected, to creating new onomatopoeic words to mimic the recorded sounds. Ultimately, I invite my students to help me divide all the sounds they heard in the recordings according to the three major categories developed by soundscape ecology: "geophony" (nonbiological ambient sounds created by wind, rain, thunder, etc.); "biophony" (sounds created by living organisms); and "anthrophony" (sounds caused by humans; Pijanowski et al. 2011, p. 204). I use the blackboard to write down their suggestions and to make the categorization immediately visible for further discussion. This takes approximately half of our lesson time.

I then divide the students into groups, taking care to split the recording pairs, and ask each group to have a conversation about both their experiences of listening and recording the soundscape while *in situ*, and their attempts to transcribe the sounds they experienced at Lyme Hill. To facilitate their conversation, I give them a few preliminary questions:

- What did you learn from listening to the soundscape at Lyme Hill while you were there?
- What sounds did you notice the most while listening to your own recording?
- How did you select and assemble the sounds for your recording sample and why?
- What challenges did you encounter in transcribing the soundscape into a written form?

As the prompts suggest, I want my students to reflect upon the following:

- The importance of soundscape for our own sensory awareness of the nonhuman world.
- The technological medium involved in the recording and the editing practices that come with the medium itself.
- The potential discrepancies between written and oral expressive forms as well as between human and nonhuman semiosis.

Finally, I ask each group of students to come up with a short report for the whole class about their conversation. The lesson ends with a final discussion about the importance of soundscape ecology, the relationships between human and nonhuman semiosis, and, more generally, what they have learned from the activity at Lyme Hill.

Conclusions

For their final projects, students develop an essay (either written or video) or a creative project crafting a "conversation" among four different sources: one of the Italian environmental texts we read throughout the whole term; one of the secondary, more theoretical readings we explored together; an environmental issue that impacts their community (either nationally or locally); and themselves, i.e., their own thinking, emotions, and embodiment. I encourage students to reuse the object, topic, or experience they introduced in either one of two short papers that they wrote in week 3 and 6, or they can introduce a new topic, but it must be a single object or experience, seen from different perspectives, rather than a broad class of objects or experiences. Although I do not explicitly suggest writing about our field activities, a few students decided to use their final project as an opportunity to reflect further on what experiencing the soundscape of a physical environment outside the classroom meant for them. For instance, one student pointed out his surprise in listening to the pervasive presence of car noises in all the recordings made by the class at Lyme Hill. This experience made him reflect upon acoustic pollution caused by humans: ultimately, he argued that not only what he called "technophonies" (any sound created by a machine or engine or other human technology) "affect the way that we perceive our environment," but the very way in which "we interact with and view nature is negatively affected by the overwhelming presence of technophonies and other human-generated noise pollution." Another student called instead the experience of listening to the soundscape at Lyme Hill "deeply transformative." As he wrote:

> Recording, editing, and describing the sounds of the wetlands, not only increased my attentiveness to sounds, but also improved my comprehension of the concept of soundscapes. I stopped envisioning my auditory environment as individual sounds, but rather one cohesive soundscape. Additionally, I now found it easier to describe different sounds as both one cohesive sound and individual parts. As the week went on, my knowledge and comprehension of soundscapes only continued to improve and my attentiveness to sounds continued [to] increase.

As these few testimonies highlight, our experience with the Lyme Hill soundscape had two major consequences for my students. On the one hand, their new soundscape awareness appears to have changed their personal relationship with their surrounding environment, increasing their overall ecological consciousness. Grasping the significance of silence and of listening to how "birds and leaves and cars and footsteps came together to form a sonic fingerprint of this location," as pointed out by another student, was a crucial outcome of our activity. In fact, during in-class conversations, several students noted how our study of soundscapes has not only changed the way in which they experience their physical surroundings but also increased their attentiveness to environmental and ecological destruction.

For instance, they commented upon using sounds to experience a more holistic view of the ecological happenings around us as well as about noise pollution as another example of "slow violence" impacting both the human and nonhuman world. One student even created a sound map of our campus, exploring the common disconnect in sound awareness experienced by Dartmouth students and noting how "increasing our appreciation towards the soundscapes we live in has the potential to make us feel more grounded—both spatially and temporally." On the other hand, and somehow consequently, students reported how at Lyme Hill and throughout the whole week devoted to "sounds," they started to realize how soundscapes can develop a sense of place unique from an exclusively visual apprehension or a mere abstract, or intellectual, understanding. Instead, a richer and multisensory comprehension of their own embodied presence within a multispecies dwelling grew out of our field activity. Students were struck by the at times uncanny connection between (their own) emotions and soundscapes, as if sounds can somehow "touch" you more than any visual cue. For example, one student shared with the class how listening to the sounds of goats in *The Four Times* triggered such a strong emotional reaction as to move her to tears, highlighting what she called "the intimate hold that sound can have over us—what it does for us in its presence and what might be imbricated in its absence."

Being exposed to the "infinite choir" of nonhuman voices that compose the soundscape of Lyme Hill—to return to the Agamben's description—forced my students to "try to speak, to think." However, this time, their words and thoughts were not the usual assertion of human centrality but the humbler attempt to engage with the world, including their own inhabiting in the world, in a more attentive and responsible way.

Acknowledgment

Indigenous people have cared for this land for centuries. The land that the Upper Valley Land Trust owns, conserves, and works on, and the land on which we all live, was first stewarded and cared for by Indigenous people of the Abenaki Nation, a tribe of the Wabanaki Confederacy. These are the traditional, ancestral, unceded lands of Abenaki people, taken from them by violence. Current day non-Indigenous people have benefited from that violence, and that is a history that we are reckoning with. Indigenous people are not gone, they live here and are a part of the past, present, and future of our land and our communities. We know this acknowledgment is a small step in a bigger process of greater awareness of Native sovereignty and cultural rights. We honor with gratitude the land itself and the Abenaki people, past and present.

Notes

1 The epilogue appears only partially in the English translation of this work (Agamben 1991). The translation of this initial passage is therefore mine.

2 In a previous week devoted to the theme of "Animal," we discuss the link between animal bodies and writing, as we point out how parchment, a material made of untanned skins of nonhuman animals, has been used as a writing medium for centuries.

References

Agamben, G., 1982. *Il linguaggio e la morte: un seminario sul luogo della negatività*. Bologna: Einaudi.

Agamben, G., 1991. *Language and death*. Minneapolis, MN: University of Minnesota Press.

Agamben, G., 2018 *What is philosophy?* Stanford, CA: Stanford University Press.

Benvegnù, D., 2019. Creative writing and art: Toward an ecopoetics of randomness and design. *Ecozon@*, 10 (1), 196–200.

Eppelsheimer, N., Küchler, U., and Melin, C., 2014. Claiming the language ecotone: Translinguality, resilience, and the environmental humanities. *Resilience: A Journal of the Environmental Humanities*, 1 (3), 54–68.

Feld, S., 1996. Waterfall of song: An acoustemology of place resounding in Bosavi, Papua New Guinea. In K. Basso and S. Feld, eds. *Senses of place*. Santa Fe, NM: School of American Research Press, 91–135.

Gallagher, M., *et al.*, 2017. Listening differently: A pedagogy for expanded listening. *British Educational Research Journal*, 43 (6), 1246–1265.

Kull, K., 1998. Semiotic ecology: Different natures in the semiosphere. *Sign Systems Studies*, 26, 344–371.

Murray Schafer, R., 1992. *A sound education: 100 exercises in listening and sound-making*. Indian River, ON: Arcana Editions.

Past, E., *Italian Ecocinema: Beyond the human*. Bloomington, IN: Indiana University Press.

Pijanowski, B., *et al.*, 2011. Soundscape ecology: The science of sound in the landscape. *BioScience*, 61 (3), 203–216.

Pratt, M. L., 1991. Arts of the contact zone. *Profession*, 1991, 33–40.

Springgay, S., 2011. "The Chinatown foray" as sensational pedagogy. *Curriculum Inquiry*, 41 (5), 636–656.

14

TEACHING ANIMAL TEXTS

American Environmental Literature's Ability to Connect Students to Animals and Wildlife through Observation

Lauren E. Perry

My environmental literature course, titled, "Animal Texts: Critical Animal Concepts for the Anthropocene in American Literature," examines how key environmental texts from the late nineteenth, twentieth, and twenty-first centuries portray animals and the changing conception of animal lives. This course provides students with an entry point into environmental literature and for them to begin reading nonhuman subjects as worthy of study, notice, and as critical elements of larger environmental projects. My goal in this course is to teach students to identify and isolate their own conceptions of what these historically significant authors were communicating about nonhuman animals and the environment. The assignments I have built within this course require that students seek out wilderness and wildlife observations, building on their experiential knowledge, and developing a deeper connection to environmental literature and the wildlife it so often portrays. The experiential and textual aspects of this course are both essential and complementary elements that encourage students to "see" animals with new eyes for how other environmentalists have seen them before. They read pivotal pieces of literature for their perspective on wildlife, and then seek out their own experiences and observations of wildlife. Combining the two allows a reshaping of students' existence as players in areas where wildlife exists.

The observational element of this course invites students to rethink the idea that wildlife is far away from them and that human spaces somehow erase or negate wildlife. As Arran Stibbe writes in *Animals Erased: Discourse, Ecology, and Reconnection with the Natural World*,

> When animals are erased, what we are left with are signs: words, pictures, toys, specimens, beeps on a radio receiver. Although the signs emerge at first with a connection to real animals, they can take on a life of their own in a simulated

DOI: 10.4324/9781003221807-14

world, becoming what Jean Baudrillard (1994) calls "simulacra"—copies without an original.

(Stibbe 2012, p. 2)

The erasure Stibbe references is the mass extinction and eradication of nonhuman animals from humans' daily lives, but the signs that convey animal life remain abundant, as is the case with environmental literature. As students know, nonhuman animals are more visible than ever thanks to social media, activist groups, and the emergence of Animal Texts like Dan Flores's *Coyote America* (2016), Nate Blakeslee's *American Wolf* (2017), Steve Rinella's *American Buffalo* (2008), and Jon Mooallem's *Wild Ones* (2013), to name just a few. Thanks to their observational journals, students also bear witness to the fact that animals are everywhere, including in the areas near where they live. Stibbe's assertion that signs in language can create a powerful connection to animals illustrates the necessity of a literary animal discourse that retains meaning rather than erases it. Allowing students to make these connections and the invaluable perspectives that they bring to the texts helps illuminate for them the power of environmental writing. Literature and the power of language are what keep animals in writing from becoming simulacra, but close analysis is necessary to understand that writing and how it has developed across literary texts. Further comparative analysis of animals, ideally by the students in this course, in literature will result in the emergence of more terms and more nuances in language that can be redefined or broken open to encompass nonhuman animals in writing more fully.

One of the first assignments of the course is an observational journal. The assignment asks students to go out into an open space (areas of wilderness here in Albuquerque, New Mexico), up into the foothills, or all the way into the Sandia Mountains if they choose. I ask students to sit for a minimum of one hour and observe any and all wildlife that crosses their path. This assignment, prior to any of our readings, encourages students to listen, see, and become intently observant of the environment around them. It also begins their journey into experiential knowledge accompanied by animal-driven environmental texts. Wildlife is everywhere, but they must sit still and observe if they are to see it. Oftentimes, the wildlife that students end up observing is not what they expect in that it might be insects on the ground or foliage, or simply listening to bird calls. I maintain that, in these spaces of wilderness, there is always wildlife present, but we must readjust our perspective to notice it.

The course readings move chronologically across the late nineteenth-century to the present day. I explain to students the importance of both wilderness observation journals and reading response writings throughout the course, as these are the primary ways I learn their writing styles and perspectives. The practice of writing and analyzing texts must be externalized, and students learn to produce quality, pointed responses to each text we read throughout the course. The parallel of having students seek out a wilderness space and journal about what they see, hear, and observe with a written response to each reading allows for a catalog of growth in both external experiences in nature and in how they are thinking about environmental readings. Weekly writing responses to what they read are integral to students' reflection on and discussion of complex concepts about the environment, especially when applying it to nonhuman animals as subjects in a way they most likely have not done previously. Shifting the focus to nonhuman animals as part of their literary world allows for a more lived-learning approach to critically important environmental texts. Response writing helps solidify their internal reactions, connections, and questions about what they read and think. Students submit response writings weekly, and I grade them on a pass-fail basis. If students attempt a thoughtful, non-value-based response, they get all points. I want to read what students are thinking and establish that their perspectives are valid and critical to the outcome of the course as a group learning environment.

The first unit of study begins with short stories by Sarah Orne Jewett and Jack London. In "A White Heron" (1886) and "To Build a Fire" (1902), I task students with examining how early environmentally minded writers developed nonhuman animals' independent subjectivity. Along with these texts, students develop an observation journal where they observe both birds and dogs. The second unit examines how Aldo Leopold's *A Sand County Almanac* (1949) and Rachel Carson's *Silent Spring* (1962) promote ecological awareness by paying attention to animal time. This observation journal requires them to observe wildlife in a city area. What wildlife can they observe downtown? How does it jive with the urban surroundings? Unit three examines Edward Abbey's *Desert Solitaire* (1968) for how it develops a layered understanding of animal consciousness. This observation journal requires them to observe a domestic animal. What kinds of clues about consciousness or conscious actions can the student detect? Unit four examines Terry Tempest Williams's *Refuge* (1991) for how it merges the genre of memoir with scientific writing to chronicle nonhuman animal memories. This observational journal is one of reflection; what was their first memory of an animal? Unit five looks at Dan Flores's *Coyote America* (2016) and Nate Blakeslee's *American Wolf* (2017) as examples of animal texts that utilize history, mythology, science, and decades of wildlife watching to create a new kind of literary animal presence that accurately conveys what nonhuman animals have experienced and continue to experience alongside humans. This observational journal requires them to think of an animal that has been historically maligned and observe it (even if only via YouTube) to see what is true versus what they have been taught to believe. Including as many students as possible in this discussion of animals in literature stands to change our cultural approach to animals ontologically. The experiential knowledge component here is vital to their understanding of these texts as they stand to represent nonhuman animals in the world beyond literature. Each observational journal pulls the students outside of the classroom and beyond the written texts to where animals live.

The animal concepts and terms in this course of study find their importance in the "animal" distancing them from their assumed anthropocentric usage with language. Through their field observations in this course, students are tasked with coming up with their own language to describe what they see and what they witness in the nonhuman animal world. Without the animal, each of those terms (subject, time, consciousness, memory, text) are presumed to be uniquely human-centered concepts. Together, the students and I dismantle this notion as we move through the texts and discuss what they observe and encounter when they sit and observe outside and in wilderness spaces. Literature can help prove that animals are players in larger conceptions of life and the world we live in. Language has leeway to cover nonhumans within the space between roots of words and living definitions. As Vicki Hearne (1991) details, our conceptions of animal "happiness" and even animal "talent" show how marked language is by how terms are imposed on animals from human perspectives. In her discussion of animal fulfillment, she writes, "I bring up these words also because they cannot be found in the lexicon of the animal rights movement. This absence accounts for the uneasiness toward the movement" (Hearne 1991, p. 60). In the nearly thirty years since Hearne pointed out the lack of understanding of words as they relate to nonhuman animals, the use of language still struggles to gain traction in bridging human concepts with animal lives. Discourses are living entities in that they change with time and culture, and they adapt to suit the contemporary moment. Through their own experiences, students create their own moments of animal connections, even if small, that create important connections to other forms of life. Having students who have lived in Albuquerque most of their lives witness coyotes in the foothills open spaces or see a porcupine walking on the Bosque trail (on the Aldo Leopold stretch, no less) help illustrate to the students the possibility of connections they might have with wildlife. Wildlife is not gone, dead, or missing; it is just being overlooked. Experiential assignments that force students out into fresh air, the outdoors, and task them with simply observing yield the most astounding results. Students observe centipedes, ants, listen to bird calls on the Rio Grande, or sometimes take note of the deafening silence that Rachel Carson must have meant when writing *Silent Spring*.

Literary animals show that writers have seen what science now backs up: that nonhuman animals and their world require complex discourse to describe them. Cary Wolfe (2003, p. 48) says that "shared language of animal training," as evidenced by Hearne's work with dogs and horses, "makes possible a common world between beings with vastly different phenomenologies." Literary studies needs to build its own animal-based shared language. Science is finally casting off the "millstone" hung around its neck by apprehension to interrogate nonhuman animal emotion, psychology, and capacity (de Waal 2016), and in response, we must view language of the past differently. Critical scholarship requires language to help it complicate definitions and terms to build a more apt animal discourse. Students begin building and experimenting with this type of language as they both read about nonhuman animals from the perspective of the authors of our course texts, and for how they use their own language to construct field experiences and memories about animals.

My students must question language in its application to animal discussions for productive literary studies but also because of the prevalence of writing about nonhuman animals in these teachable settings. Language can help disseminate scientific findings and the evolution of thought pertaining to animals. During my first semester of teaching English composition at the University of Wyoming, a student cited in her final paper that one reason endangered species laws should not protect wolves was because of the brutal way a wolf had slaughtered her family's livestock. Her specific rebuke was that the wolf's method and motivation for killing were "inhumane" in that the wolf had killed not to feed but merely for sport. My immediate reaction was that, of course, the wolf's method was not humane as the wolf is not human. I assumed that the etymology of the adjective found its root in the word "human." The *Oxford English Dictionary* does describe "humane" as originating as a variant of the word "human," but the word's first listed definition is actually "civil, courteous, or obliging towards others (*obsolete*). In later use: characterized by sympathy with and consideration for others; feeling or showing compassion towards humans or animals; benevolent, kind." In hindsight, my knee-jerk reaction that a wolf cannot be humane nor be expected to act humanely simply because it is not human was inaccurate. Once we acknowledge its root, the definition of the word is based on actions that convey thinking, emotion, and intention. The *OED*'s subsequent definition of the word is "Designed or calculated to inflict minimal pain" which is quite different from the first. From texts like *American Wolf* (2017) and entities like The Yellowstone Wolf project, we know that wolves can and do calculate, design (in their own ways), and are cognizant of pain in both playing and hunting. The more one learns about wolves, the more one sees that "civil, courteous, and obliging towards others" is exactly what could describe particular pack dynamics.

My point in encouraging students to question the definition of the word "humane," especially in conjunction with human fear and malice towards other apex predators, of which there are so few left, is to identify slippage in the word itself and how it applies to nonhuman animals. Denotatively, wolves can be humane. That memory comes back to me because of the disconnect between the agriculturally-minded student and my idea of "humane" because I failed to question humane as a term originating in the word human. If I had investigated the term further, I would have found that it lacks ties to explicitly human traits. As Gary Francione writes:

> I maintain that we ought to abolish animal use altogether and not seek to regulate our exploitation of animals to make it more "humane." … I maintain that we have no moral justification for treating animals as replaceable resources—as our property—however "humanely" we may treat them or kill them.
>
> *(Francione 2008, p. xiii)*

By placing the word *humane* in quotes, Francione calls into question its connotative usage in reference to animals. His work examines animals' rights as sentient beings, and he redefines "personhood" as applicable to animals. Francione encourages his reader to examine the implications of discourse that often surround the collective, societal treatment and writing of animals in moments like the one above. Collaborative writing and language like the kind encouraged in this course beget thinking and acting, and scholars like Francione use gestures like this to highlight how discourse needs to be changed. A "humane" death for an animal that wants to live is unjust and not humane by definition. Discourse matters in terms of evolving language and reality.

Language has been instrumental in the construction of nonhuman animals in writing and in thinking. As Frans de Waal (2016) explains, science has only begun to understand different animals' intelligence and emotional capability by changing its approach, which requires dislodging all kinds of anthropocentric holds on thinking, practice, and logic. De Waal emphasizes the proven need to adjust scientific approaches to understanding animals: "Instead of testing animals on abilities that we are particularly good at our own species' magic wells, such as language—why not test them on their specialized skills?" (ibid., p. 22). Human limitations have stood in the way of advancement in thinking about and proving animal intelligence, not animal limitations. Not only that, but our specific "magic well" of language is not very adept if it fails to evolve along with our understanding of other living beings around us.

Our language for discussing animals requires scholarly renovation. The most promising avenues for change are the reflective study of older texts and the resulting creation of a more unified discourse for texts and scholarship going forward. Language must incorporate new scientific truths. Similar to how De Waal proclaims we must measure each species's capabilities with scientific experiments that make sense for their skills, we must study writing about each species for how individual writers were seeking to meaningfully connect language with those animals. Just as I assumed the word humane could not be applied to a wolf because it is not human, I recognize how its components can be attributed to a wolf in rethinking its definition. My course and my students' ability to reassess the definition for how it fits a wolf is the product of me reading decades worth of field notes about wolf observations and studies. Students typically respond by recognizing how human-centric language is, and understandably so, but also how that language seems to keep any positive or intelligent descriptors from becoming attributed to animals and wildlife. It allows us to consider that word for its potential departure from solely anthropocentric usage. The ability to remain true to language but also faithfully allow new associations is imperative for the future of ecocriticism and environmental writing. Scholars like De Waal, Wolfe, Jon Mooallem, and Peter Godfrey-Smith lead this charge, but the difference needed is one of literary inclusion of those texts.

Other scholars have seen a similar need for the reconsideration of language to suit evolving understandings of the environmental world by coining new terms and insisting on the power of language as part of real-world change. The next step in mining significant literature contributions to animal studies and real-life progress in human relationships with nonhuman animals is to rethink and reread key texts in light of animal discourse. In an animal discourse, words like humane can be recognized for their origin in believing humans to be the only beings capable of sympathy, benevolence, and a desire to thwart pain, but compiled knowledge of wolves, dogs, dolphins, elephants, and so many other animals reveals that animals can and do act in humane ways. Another example of language applied to animals meaning something vastly different is the use of the word "body" in Godfrey-Smith's (2016) investigation of cephalopods. Their bodies have no left nor right, can change shape to fit through spaces as small as their eyeballs, and yet their "bodies" are meant to be thought of the same as a human's. These examples help illustrate how important developing a literary animal discourse is to writing and thinking. Language is where science, history, experience, and observation can come together to redefine words or, better yet, tease out dormant meanings that perhaps were already being used by other writers. Lawrence Buell (2001, p. 31), in his argument for the importance of a toxic discourse, explains,

> There seem to be at least two reasons why the discourse of toxicity has not been treated with the same attention as its chemical, medical, social, and legal aspects. One, certainly, is the pragmatism that plays a major part in shaping agendas of public discussion. "Discourse" may seem a low priority when health or property is at risk.

Buell explains that our engagement with environmental issues has primarily been factional, so forming an aligned discourse has been challenging or hardly attempted. As he states above, the discourse of toxicity belongs in the same categories of influence as those that have seemingly more real-world categories. How we talk and write about things impacts how we think about them and how we internalize and approach them when trying to change them. Students are crucial witnesses to this truth. Students' ability to recognize weakness and flaws in language and terminology involves their minds and perspectives in its rewriting. Their ability to point out slippage in how we think and talk about animals makes the next generation of environmental thinkers and writers that much more apt to recruit more minds to the cause. An animal discourse requires the reexamination of literature from over the past century to thoroughly analyze possible contributions in the form of terms. As it grows, animal discourse within literature will showcase Humanities scholars' willingness to interrogate animal intelligence as writers documented it in texts created before science undertook those challenges.

Terms like "Anthropodenial" (de Waal 2016, p. 22) illustrate the necessity to include new vocabulary to encompass aspects of nonhuman animal existence in writing. De Waal defines Anthropodenial as the human recognition of shared emotion or motivation in animal actions but the immediate dismissal because of human exceptionalism. As de Waal (1997) wrote of his then-new term, "Endowing animals with human emotions has long been a scientific taboo. But if we don't, we risk missing something fundamental, about animals and us." This term depicts a phenomenon that frequently occurs for humans yet illustrates the vast gap in knowledge, data, and even curiosity about the truth of similarities we observe. The internet loves videos of diverse species becoming friends and playing, and the more our cultural mindset is opened to the possibilities of nonhuman animals living complex lives, the more necessary a working, shared animal discourse becomes. While at first, that discourse might serve to join and streamline academic discussions of animals across literary studies, animal studies, science, philosophy, and the like, its usage in more accessible settings like environmental activist groups might help larger populations in the process of rethinking animals. New, scholarly vocabulary serves to establish the credibility of progressive ideas about animals, which impacts animal rhetoric beyond academia. The assumptions most people make when visiting the zoo, watching a documentary, or witnessing a wild animal out in real life might be corrected if more emphasis was placed on how we speak and write about animals.

An animal discourse in literature already exists in a limited capacity. In my course's inquiry into a few of environmental literature's most important texts, literary animals demonstrate writing's ability to convey sentiments that deviate from societal norms. The students in my class bring their own spin on animal discourse as they seek to recapture their own nonhuman animal experiences and to convey what they have witnessed in the animal world. Observing birds that impressively flock together as a unit over the bosque along the Rio Grande and coyotes that communicate through song and interactions demonstrate an animal world and cognizance that students are keen to analyze. American writers whose views of the continent and the world changed cultural perceptions of the environment also included animals in their writings, and their use of language surrounding nonhuman animals requires close-readings and analysis. Those close readings and analysis then come to life alongside personal experience in nature and the specificity of observed animal actions in a space that is not far from students' homes. Experiential knowledge about how special, unique, and fragile each ecosystem and wildlife encounter is providing students with tools of animal awareness that helps them infuse literature with personal connection. In studying this duality of language in writing about animals, many environmental texts stand to bolster and contribute to that discourse. As Cary Wolfe (2003) notes, discourse has an immense impact on how concepts from literature are viewed in real-world scenarios, especially for animals and others. There are yet so many texts about animals and the environment that we have yet to examine the nature of the writing used to discuss them and convey their existence. Aaron Gross writes:

> Animals matter. Animals matter not only because they live, but have lives—lives that we have every reason to believe are filled with joy and pain. But the way animal lives matter is always complicated by how we use animals to shape the landscape of our humanity, both materially and imaginatively.
>
> *(Gross and Vallely 2012, p. ix)*

Language is a significant contribution to both the material and the imaginative manifestation of that humanity. Other terminology might be created, like *anthropodenial*, or terms may be reclaimed for animals within literature that have yet to be mapped out.

This type of course, with a literary inquiry into animals paired with the encouragement to strike out into field observations, and others like it, are helping to de-anthropocentricize language itself. My student may not have known that she could call a wolf inhumane, but perhaps she intuited that other nonhuman animals she has interacted with are and have the potential to act humanely. Like the internet video of a German Shepherd gently nudging cheerios out of a baby's hand, animals are aware of the world around them. We, too, are aware of them in ways that literature and language might have always already been conveying. By investigating, utilizing, and cultivating our application of animal terms to develop an animal discourse, we make it easier for writers and scholars alike to produce innovative literature about animals. The goal should be to analyze and learn from nonhuman animals in literature of the past and to trace how it informs how we speak, write, and collectively see animals at present. I think if studies of animals within literature continue in specific, critical ways, we will discover how much language has evolved around nonhuman animals the way animals themselves evolve. Just because humans create language does not mean there is not a place for animality and nonhuman creatures within terminology, concepts, and theoretical frameworks that help us make sense of the world.

Across literary history and the shelves of North American animal writing, there are so many subgenres and powerful breeds of texts. Imagine what new terms and additions to a literary animal discourse might arise from this course and its analysis of texts that chronicle the journeys of animals into new spaces and, vice versa, the retreat of humans into the territory of nonhuman animals to escape the human. Gavin Van Horn (2018) reclaims terms like Inhabitation and Anima in structuring his narrative while also incorporating Spanish to detail his investigation into urban coyote lives. Doug Peacock's publications on his life among grizzly bears (e.g. Peacock 2013) have inspired throngs of people to protect the threatened existence of one of North America's great bipedal predators. Ideally, the sustained investigation into writing about animals, whether in environmental writing, literary classics, or cross-disciplinary texts, would allow for the development of a recognized animal discourse, a reexamination of the language we use to write animals, and the potential for grouping narratives about animals so that scholars might construct meaningful patterns from past observations across many forms of literature. London, Jewett, Carson, Leopold, Abbey, Williams, Flores, and Blakeslee have illuminated conceptual continuities within writing about animals, and I am confident that the study of other environmental texts by eager students will reveal further duality in language and connections in the thinking of writers as they attempt to write animals.

The final assignment of the course is an argumentative research paper and technological presentation in which each student creates their own animal concept. Their semesters' worth of observational journals, both of wilderness, wildlife, domestic animals, and their own animal memories help equip them to create their own addition to the animal discourse. Students create this animal concept and term by building it out from their close readings of whichever course texts they choose, as well as informing it with cross-disciplinary scholarship related to the concept.

Their observations are the key to helping them choose what stands out to them about the nonhuman animal world, and what connections they have made between what they have read, witnessed, and thought about as they have pondered animals in literature and life. The student must prove the concept to exist in literature about animals, and the animal concept must enhance literary understanding of the real presence and lives of animals. Their presentation to the class is one that conveys the main proofs of the concept through textual analysis and also how that term stands to benefit real animals through activism, understanding, and ontological change. As with the response writings and observational journals, this assignment requires students to pull from their own experiential knowledge and research to develop a concept that connects to the reality of environmental issues concerning and threatening nonhuman animals.

Acknowledgment

The University of New Mexico in Albuquerque, New Mexico, sits on the traditional homelands of the Pueblo of Sandia. There are over twenty-five tribal nations who have territorial connections to the State of New Mexico, and I want to acknowledge all of them, especially the pueblos in the area of Albuquerque.

References

Buell, L., 2001. *Writing for an endangered world: Literature, culture, and environment in the U.S. and beyond*. Cambridge, MA: Belknap Press of Harvard University Press.

Blakeslee, N., 2017. *American wolf: A true story of survival and obsession in the West*. New York: Crown Publishing Group.

de Waal, F., 1997. Are We in Anthropodenial? Retrieved from www.discovermagazine.com/planet-earth/are-we-in-anthropodenial (accessed April 10, 2021).

de Waal, F., 2016. *Are we smart enough to know how smart animals are?*New York: W. W. Norton & Company.

Flores, D., 2016. *Coyote America: A natural and supernatural history*. New York: Hachette.

Francione, G.L., 2008. *Animals as persons: Essays on the abolition of animal exploitation*. New York: Columbia University Press.

Godfrey-Smith, P., 2016. *Other minds: The octopus, the sea, and the deep origins of consciousness*. New York: Farrar, Strauss, and Giroux.

Godfrey-Smith, P., 2017. The mind of an octopus. *Scientific American Mind*, 28 (1), 62–69.

Gross, A. S., and Vallely, A., 2012. *Animals and the human imagination: A companion to animal studies*. New York: Columbia University Press.

Hearne, V., 1991. What's wrong with animal rights: Of hounds, horses, and Jeffersonian happiness. *Harper's Magazine*, 283 (1696), 59–64.

Mooallem, J., 2013. *Wild ones: A sometimes dismaying, weirdly reassuring story about looking at people looking at animals in America*. New York: Penguin.

Peacock, D., 2013. *Grizzly years: In search of the American wilderness*. New York: Holt.

Reed, A.E., et al., 2020. Companion animals and online discourse: Victim-blaming and animal evacuation. *Anthrozoos*, 33 (6), 727–742.

Rinella, S., 2008. *American buffalo: In search of a lost icon*. New York: Spiegel & Grau.

Scollen, R. J., and Mason, A., 2020. Sea World—Gold Coast, Australia's discourse of legitimation: Signage and live animal shows (2015–2018) as indicators of change in Messaging. *Journal of Sustainable Tourism*, 28 (10), 1686–1701.

Stibbe, A., 2012. *Animals erased: Discourse, ecology, and reconnection with the natural world*. Middletown, CT: Wesleyan University Press.

Van Horn, G., 2018. *The way of the coyote Shared journeys in the urban wild*. Chicago, IL: University of Illinois Press.

Wolfe, C., 2003. *Animal rites: American culture, the discourse of species, and posthumanist theory*. Chicago, IL: University of Chicago Press.

15

TO THE ZOO!

Jeremy Chow

I like to think of myself as an environmental humanist. Teaching the environmental humanities requires a commitment to true interdisciplinary work: pedagogies that enfold humanist inquiry and methodologies alongside topics, studies, and discourses that have previously been conducted under the aegis of biological or environmental sciences. In classes such as "Literature and the Sea," "Animal, Plant, Human," "Animal Stories," and "Chimp Lit"—on which I will focus at length here—I remind students that work on the environment, more generally, and nonhuman animals, more specifically, is too important to be territorialized. That is, environment and animality as material and theoretical inquiries cannot belong to only one field. As Neimanis, Åsberg, and Hedrén (2015, p. 80) cooperatively write, "any policy or action aimed at ameliorating environment problems must take into account human desire, motivation, and values; a deep understanding of environment cannot be divorced from human imagination, culture, and institutional and social practices." The three argue that one of the many values of the environmental humanities lies in mitigating disciplinary monocultures. In specific, my nonhuman animal-attentive courses enable students to embrace the environmental humanities, for it has much to teach us about how we read, understand, and interact with nonhuman animals. The field of "more-than-human history," O'Gorman and Gaynor (2020, p. 2) posit, "seeks to challenge human exceptionalism, instead situating humans as participants in manifold ecologies, with histories of, and possibilities for, becoming-with nonhuman beings shaped by our changing, diverse, and entangled lives." Such a theoretical and practical attention fosters alternative, nontraditional, and interdisciplinary approaches to grappling with the question of the nonhuman animal.

This chapter explores how a literature curriculum can evolve with the benefits of experiential and experimental pedagogies that invite students out of the comfort zone of the walled classroom to encounter a living, learning classroom that resides in nonhuman animal-focused institutions. My students and I regularly visit

DOI: 10.4324/9781003221807-15

nonhuman animal research labs, zoos, and sanctuaries, which allow us to reframe the two-dimensional readings by welcoming the three-dimensional, lived realities of these institutions. These institutions are not without their appropriate criticisms, and the type of educational tourism my students and I participate in also has rightful criticisms that we air together.

My students and I recognize our complicity in systems of animal capture and containment—for research and entertainment purposes—yet, I stand by the decision to allow students to approach these institutions as sites of critical inquiry, rather than exclusively complicit consumers of conspicuous animal consumption. For example, we actively discuss the role of animal captivity in the popular imaginary and the repeated shortcomings of conservation missions. A humanistic eye for critical inquiry thus allows students to approach a variety of "texts" that include literature, but also art, film, media, music, advertisements, and institutions of all sorts. I promise my students, then, revived ways of seeing, interacting, and thinking, which demonstrates the vitality and necessity of the (environmental) humanities. Human–animal relations and nonhuman animal-centered institutions make that promise easy to keep. I offer "To the Zoo!" as a pedagogical commitment to diversifying curriculum, breaking down disciplinary silos in the academy, and lionizing the humanities as germane to apprehending the nonhuman animal and its placement in our immediate and peripheral concerns.

Chimp Lit

I regularly teach a course, "Chimp Lit," that explores the relationships among humans and nonhuman simians within literature, contemporary culture, philosophy, and psychology. A rather unusual offering (albeit one that is typically over enrolled), "Chimp Lit" is how I fine tune the literature and environment curriculum. Typical literature and environment courses, while increasingly popular, favor longer surveys of environmental thinking from antiquity to the present, though repeatedly focalize American environmentalism, as evidenced by the Association for the Study of Literature and the Environment (2021) pedagogy section. "Chimp Lit" is admittedly not a survey; its topical specificity appeals mostly to students outside of the humanities, especially those interested in biology, environmental studies, and animal behavior. "Chimp Lit" zeros in as a case study that allows for extrapolation, distant and close reading, and interdisciplinary metacognition.

Students are generally unfamiliar with the sheer number of primates that populate beloved fictions and media. We read short stories from Japanese folklore, Haruki Murakami, Edgar Allan Poe, H. P. Lovecraft, and Franz Kafka. We weep over novels such as Karen Joy Fowler's *We Are All Completely Besides Ourselves* and Colin McAdam's *A Beautiful Truth*. We explore the simian in the filmic imaginary with blockbusters like *King Kong* and *Planet of the Apes*. We attempt to make sense of extractive economies that demonstrate the precarity of primates and the independence of African nations, such as the Democratic Republic of Congo and Rwanda, in documentaries like *Virunga*. We approach graphic narratives and comics such as Marvel's *Tarzan*, Gene Luen Yang's *American Born Chinese*, and Jim Ottaviani and Maris Wick's *Primates: The Fearless Science of Jane Goodall, Dian Fossey, and Biruté Galdikas.*

Primates serves as a gateway to popular primatological journalism that introduces students to nonfiction, psychology, ethology, and philosophy; that is, on the heels of Ottaviani and Wick, we read Goodall, Fossey, and Galdikas in their own words. In so doing, we open redefined worlds of what constitutes "literature," and likewise ease students into non-traditional literatures through a genre (the graphic novel) that is highly accessible. Scientific non-fiction and autobiography are genres limitedly taught in most English departments because they are often disallowed from consideration of "great literature." My students and I are repeatedly struck by how Goodall, Fossey, and Galdikas employ first-person narratives to demonstrate their situated knowledges and rail against the ostensible "objectivity" of scientific writing that would otherwise censure the research these women establish. Like most universities now, mine enrolls more women students, and many are already familiar with experiences of systemic and institutional intersectional misogyny. As we discuss in class, it is one thing to experience the loneliness and imposter syndrome that ensues from these experiences that trickle down in class, marginal comments, or standardized microaggressions; it is a wholly other thing to encounter these topics in our readings, rife for discussion. Goodall, Fossey, and Galdikas's diaristic prose is replete with humor, horror, and rhetorical pathos: the three write of personal hardships, death of human and primate kin, systemic chauvinism, and discoveries that shock and infuriate the scientific world. Of their simian interlocutors, they document sexual violence, cannibalism, generational trauma, predation and poaching, group politics, and affinities for emotional attachment and attunement. The documentation that the three provide models ethological observation. In other words, they participate in a commitment to observation of the nonhuman animal as a means of understanding and interpreting animal behavior, the need for conservation, and the unique capabilities of nonhuman species. Since the class favors analysis and interpretation, Goodall, Fossey, and Galdikas articulate an implicit form of argumentation, which I task students with replicating. This is what prompts my students and I to close our books, exit the class, and journey off campus.

Following our reading and discussions of *Through a Window, Gorillas in the Mist,* and *Reflections of Eden* my students complete an assignment I call the "nonhuman observation and synthesis." Students are invited to approach a nonhuman animal as a "text" that can be read, interpreted, and argued, just as we've witnessed in the feminist-oriented primatological narratives. This is, I acknowledge, a strange task that doesn't attempt to elide all texts but rather seeks to apply forms of reading, interpretation, and argumentation that are informed by and yet live outside of literary traditions. Such an assignment makes clear that analysis and close readings are humanistic methodologies that have wide applicability; they represent a form of visuality and knowledge. How do we animate modes of reading that make available the wide breadth of humanistic inquiry? This assignment attempts to embolden such strategies.

We approach this activity at the zoo or other Association of Zoos and Aquarium (AZA)-accredited institutions: in the inaugural iteration, at the Santa Barbara Zoo, and in other iterations, Reptiland in Allenwood, Pennsylvania. These respective field trips allow students to explore the backyards of our distinct collegiate environments to foster relationships among institutions and to visualize research and networking opportunities outside of the university.

When we arrive at the zoo, students select a particular species, either an individual or group within a species, and document "field notes" for a minimum of thirty minutes. These field notes take on any number of forms: bullets, sentences, video captures, or screen shots. They thus emulate the types of narrative field notes that the three primatologists reveal in autobiographies. I do not require that students standardize their field notes; my students approach notetaking from different trajectories, experiences, and learning styles. Field notes thus become a way of honoring learning difference in the classroom all the while seeking to model ourselves after Goodall, Fossey, and Galdikas. Thirty minutes is still insufficient, but it is almost 400 to 1500 percent more time than the average zoo goer spends in front of a single animal exhibit (Marcellini and Jenssen 1988; Ross and Lukas 2005).[1]

At the end of the visit with field notes in hand, students return to the classroom to interpret their data and craft an argument about animal behavior based on their observations. The field notes act as a guide for their argumentation. "Think of these notes as your evidence," I tell my students. With this evidence in front of them, students seek to pattern, elucidate, and make sense of what they've observed of captive nonhuman animals. I acknowledge that these field notes do not make them de facto experts; they are not suddenly animal behaviorists. Goodall, Fossey, and Galdikas have dedicated their lives to these types of observations and work. Thirty minutes pales in comparison. However, even the first-hand experience that this assignment provides lays the groundwork for thinking about interpretation and making meaning.

Lessons from the Santa Barbara Zoo

The Santa Barbara Zoo welcomes my students and me on a characteristically cloudless Santa Barbara day. Prior to our visit, I have arranged for the class to interact with staff who generously offer their time to discuss the zoo's history and its current holdings. This requires a fair amount of legwork on my part. It includes a variety of email exchanges with the zoo, my department chair, and staff that manage departmental finances. In some instances, the site (eager to collaborate with college students and potential volunteers) offers free admission; in others, admission is discounted. While excursions like these are minimized in the humanities, universities, in my experience, pride themselves on their relationships with (or ingratiate themselves to) the local communities. Emphasizing the community-based knowledge systems that are constructed through experiential learning is often an easy rationale that gets checks written.

At the Santa Barbara Zoo, the megafauna, without question, attract the students, and this accords a pattern noted by other zoo researchers and environmentalists.

Such a realization also endorses the zoo's desire and justification to house additional megafauna. The Santa Barbara Zoo is home to (at that point) six Masai giraffes, and the most reproductively valuable giraffe sire in the US, all of whom are shown in Figure 15.1. The Masai giraffes are, in many ways, the crown jewel of the zoo. Our docent escorts us towards the giraffe enclosure, which looks out towards the Pacific Ocean. The giraffes strut beneath us. The enclosure is structured to accommodate an a la carte option popular at many institutions that house giraffes: feeding, which costs an additional $10–15. In this way, the structure is built to make possible these a la carte experiences and thus foster up-close-and-personal encounters among humans and nonhumans. Because such experiences are now quotidian at zoos, my students accept them without question. When we return to the classroom, in discussion, I invite my students to linger over the politics and ethics of these experiences. How, for example, are animals further commodified by these experiences? And why are they so attractive to audiences? Questions such as

FIGURE 15.1 A tower of giraffes at the Santa Barbara Zoo (2017).
Source: Jeremy Chow.

these invite students to consider the limits of conservation missions, the attraction of nonhuman animal sites, and the ways nonhuman animal bodies become commodified for entertainment economies.

I further press my students to consider the infrastructure of these exhibits. Through our guide, we learn that giraffe horns grow superficial tumors *only in captivity*. The reasons are unknown. We learn that all nonhuman animals in the zoo are on some form of birth control. "All" means only females of the species. We learn that the AZA regulates which nonhuman animals may reproduce and which cannot. We learn that male giraffes will sample the urine of females to ascertain whether the female is in estrus; we watch this happen several times. Parents shy their children away from this spectacle. We learn that the two young male giraffes (sired by the resident male) will be "loaned" to other zoos shortly because they will otherwise risk giraffe in-fighting (with their father) and social exclusion (by the tower) as they mature. Because they are reproductively valuable to the AZA's sire book, they cannot be harmed: they are the future of the Masai giraffe breeding population in North America. My students furiously jot down notes.

A stone's throw from the giraffe enclosure, we find two brother gorillas. This grouping attempts to position species of similar geographies together, for example, meerkats, giraffes, gorillas, fennec foxes, and lions represent African fauna (if such a thing is possible given the size of Africa as a continent and its various biomes). We pause to observe the brothers because our guide is their former keeper. Our guide conceals herself because, she tells us, the gorillas must not see her. The brothers have recently begun training with a new keeper, and the acclimation is going poorly. Gorillas in captivity do not like change. Their trust has been betrayed. If they see her, they will either refuse to work with the new trainer or become aggressive towards the keeper that has now betrayed them. One of the brothers sits in front of the viewing glass staring out at the audience, which is populated mostly by young children and their guardians. These children are encouraged to emulate gorilla behavior. They scream, shout, run around in frenzied ways. The gorilla that looks out at them does none of these things. The guardians and my students laugh. I have not asked my students to reflect on how the nonhuman animal observers interact with the enclosed species, but, as comes up later in class discussion, this is part and parcel of the zoo experience and animal ethnography in sites of animal enclosure. In other words, an ethnographic study done in a contemporary site of nonhuman animal enclosure must always account for how human audiences participate in and reframe experiences of observation.

Following our guided tour, students explore the zoo on their own to find the species on which they will write their "nonhuman observation and synthesis." Some students stay to complete their field notes on the giraffes or gorillas. Others seek out lesser-known species. For example, in their observations, students reveal the burrowing behaviors of meerkats. Some watch ornery goats harass ponies larger in size. Others watch penguins as they attempt to flee inside only to find those entrances bolted closed. One watches armadillos lounge underneath the heat lamps that cast a sepia color (simulating nighttime) around the tiny enclosure. Another observes a snow leopard pace back and forth as the sound of trucks and traffic from the 101 drone over the walls of the enclosure. The observations are enlightening—sometimes harrowing.

This assignment proves both exciting and difficult for students. They easily fill pages with their field notes: careful attention to behavior, enclosure objects, interactive accessories, and any contact among individuals or animal and keeper. But the assignment requires more than just observation. It demands synthesis: a higher-level mode of thinking that unites observations, readings, pre-knowledges, and emerging hypotheses. And while my students air their constructive criticisms of such an assignment (time limitations; lack of training as zoologists; years since formerly attending a zoo), what they invariably bring to life is a mode of reading and interpretation that enfolds the living, breathing nonhuman animal into modes of interpretation. The nonhuman animal is, unlike many texts we encounter, not passive or silent. Instead, the assignments prime students to promote specific claims based on multivocal expressions of evidence—and this is an exercise that they can take with them into any field.

Lessons from a Primate Lab

When I moved to central Pennsylvania from California, zoos like the Santa Barbara Zoo were unavailable. I needed to reorganize my pedagogical goals. As kismet

FIGURE 15.2 A capuchin monkey, Nova, who resides at the Bucknell Primate Lab (2019).
Source: Kailyn Carr.

though, my current institution is home to a primate lab—the only liberal arts college to house one, and one of only a handful of similar labs in North America directly connected to an institution of higher education. The Bucknell Primate Lab is home to four primate species: capuchin monkeys, squirrel monkeys, hamadryas baboons, and lion-tailed macaques. The only testing done on these groups is behavior and psychology focused. There are no intravenous or biomedical experiments. This reassures me. It also makes my decision to bring my students to the primate lab—deliberately hidden by extensive foliage on the periphery of campus—less ethically murky.

Unlike the "nonhuman observation and synthesis" that my students undertake at AZA-accredited institutions, the purpose of our visit to the Bucknell Primate Lab is more exploratory in nature. In other words, we move beyond the traditional classroom to a lab classroom to explore the lives of lab animals rather than those exclusively at zoos. The visit provides an opportunity for students to imagine a truly interdisciplinary, liberal arts undertaking: we interact with an animal behavior/ psychology professor. This is also a type of organizational labor that must be accounted for prior to the start of the semester. While most animal behaviorists are entertained or bemused by a *literature* course on animals ("I didn't even know that was possible," one blurted out to me when we first met), they are overwhelmingly in favor of hosting conversations about these sites where they are experts. To be clear, this type of collaboration does not ingratiate the humanities to the biological sciences; it instead forges pathways by which these fields can embark on a series of connected rather than divergent conversations. For example, the animal behaviorist and I hold a conversation with the students as we explore the four enclosures, and students thus benefit from a conversation that pairs faculty with distinct research styles and trainings with a "text"—once again the nonhuman animal body—of mutual interest.

As with the field trips to the zoo, the primate lab is organized by species, each of which is housed behind a locked door. Two of the four species have access to outdoor areas, but this is increasingly limited by the climate in central Pennsylvania. Most of the individuals have lived exclusively at the primate lab or have been loaned out (for breeding purposes) by collaborating zoos. Indeed, it is through questions about how the individuals came to live at Bucknell that we learn about breeding programs and reproductive value in particular species: males of the species are seen as more valuable because they can inject genetic diversity into a breeding program more regularly. This conversation accompanies an interaction with the hamadryas baboon population, which is headed up by a male baboon, Capaldi, that has sired nearly a dozen hamadryas baboons with his "harem." Capaldi makes himself known as the class observes the troop. He makes ferocious noises as he presses his body against the grid of the cage, revealing an erection. This is not a flirtatious appeal; it is a territorial display. Students are struck silent in beholding a troop of baboons six feet away, all of whom live in a dorm-like structure not unlike their own.

Whereas students may be less motivated to pose questions in a traditional classroom, it's clear from these experiences that their questions and their curiosity truly come alive when in a space that is shared with nonhuman animals. A small handful of my students are majors in animal behavior, and thus they have an intimate familiarity with not only the lab but also the individuals housed in the lab. The capuchins, for example, are the easiest to work with because they are highly food-motivated and enjoy the interaction with the keepers. Figure 15.2 showcases this behavior. There is no physical contact between researchers and animals. Communication between individuals is conducted exclusively through touchscreens and with food incentives. The screen mediates contact. The experiences in the lab break down disciplinary silos by demonstrating how this research is living and breathing through our nonhuman animal contacts. The primate lab field trip then revives research in three-dimensional ways, which holds cachet for students. The visit coincides with registration for the subsequent semester, and my students scramble to register for the thirty seats available in the animal behavior entry class, which is hosted in the lab.

Lessons from the Pandemic

In Fall 2020, in the throes of the raging pandemic, my college, like others, opted for hybrid instruction. While this iteration of "Chimp Lit" alternated between Zoom and socially distant in-person instruction, it was impossible to make site visits. The "nonhuman observation and synthesis" needed retooling. While I considered scrapping the assignment entirely, my surveilled newsfeed brought details of zoos and aquariums that had also opted for hybrid visits: through zoo cams. The technology of zoo cams is not distinctly a result of the pandemic; it is often used by zoo staff to monitor individuals considered medically vulnerable or females of species due to give birth. Zoo cams in the pandemic thus sought to ameliorate the isolation animals may have experienced with the lack of daily visitors—a conditioned reality abruptly halted by stay-at-home orders. As Anna Peele (2020, n.p.) described in *The Washington Post*, while some animals reveled in the solitary confinement of the pandemic (only intermittently interacting with keepers), others, especially social animals, were now starved for attention. Peele cites a keeper who suggests, "You think you're coming to the zoo to watch the animals. Well, the animals are watching you, too." Peele (ibid.) identifies this as animal "loneliness."

The fall 2020 version of "Chimp Lit" thus invited students to find a primate zoo cam at any institution across the globe and to access the live feed for no less than an hour. The benefit of a zoo cam is that students could log on and log off the browser several times over several days. This is precisely what I invited them to do.[2] Whereas when we visit zoos, our time is condensed, with the zoo cam, there is a *potential* for more accessibility. My students realized in locating streaming platforms at the Detroit Zoo, San Diego Zoo, Bristol Zoo, and Gorilla Rehabilitation and Conservation Education Center (GRACE) in the Democratic Republic of Congo that the types of visual observations one can make is entirely dependent on the camera's focal point. In other words, while zoo cams are popular resources, the technology is not reactive. The camera focuses on a lone point, and if the nonhuman primate is not within the aperture, then you're out of luck. More than half of my students reflected on this shortcoming.

In addition, while watching the primates over several days provided a revised way of interacting with the animal, my students found it difficult to discern whether the individuals they were observing were the same from the previous session. This forced students to analyze new forms of behavior that the original assignment had made easier. For example, many students documented what they identified as a gorilla "bully" through GRACE's live zoo cams. Because many of the gorillas housed at GRACE are young or adolescent individuals abandoned by their troop or, more commonly, orphaned due to poaching, the exotic pet trade, and environmental degradation, the secondary sex characteristics that might allow an individual to decipher biological sex were made fuzzy. Three students reported watching an individual repeatedly steal food from another. Two noted that one individual would repeatedly "pound" on a playmate. These behaviors, they learn from Fossey in particular, are naturalized play behaviors among gorilla youths.

The use of the term "bully" then comports a variety of interpretative moves. In one reading, it positions the gorilla individual as possessing a malice, intentionality, and determination to harass another. In another reading, it demonstrates that physical aggression is the sole hallmark of bullying, which we know to be patently false. In yet another, it shows that bullying is a behavior that is not distinct to elementary school playgrounds or social media used by humans alone. I'll note that my students do not closely read this word, but in its repetition, it provides an opportunity for me to engage with my students' socially constructed ideologies about behavior and attempt to make sense of them across species lines.

Another unintended consequence of the zoo cam is the lack of auditory feedback. The zoo is a phenomenological *mélange*. In addition to the visual panoply of dissimilar species only meters from one another, areas of the zoo have distinct smells, certain locations are humid while others are chilled, and the sound of particular species can echo from the enclosure to the parking lot or beyond. The zoo cam technologies available to my students focused exclusively on visual apparatuses. There was no way to sense the humidity or heat of orangutan exhibits that attempt to mimic Indonesia. There was no way to experience the pungent musk of gorillas, which Fossey notes always preceded a physical and visual encounter. There was no way of hearing the pant hoots that Goodall documents (with accompanying facial expressions) of chimpanzees who seek a call and response within their groups or outside. My students vehemently disliked the lack of sensory overload that too often results from a visit to the zoo. They wanted to hear the sounds that accompanied gorilla "bullying." They wanted to listen to orangutan males respond to orangutan infants. They wanted the phenomenological stimuli that they had enjoyed upon every previous visit to the zoo. This, though, is not singularly a failure of the zoo cam. It instead bespeaks the types of acculturation all students have encountered in their previous zoo experiences that prime them for particular expectations that the pandemic obviates. The zoo cam revision thus teaches us more about our expectations and preconceived notions of human-nonhuman animal interaction that lay the foundation for a vital reassessment of the zoo as a site of critical inquiry.

Would I teach with the zoo cams again? *Yes and no.* While I have been fortunate to collaborate with resources that make the visits financially viable (and to work with classes that are manageable as a field trip), the zoo cam provides an accessibility that requires zero financial or administrative support. It is, then, easily adaptable to any type of program or class. To be truly effective, though, I feel that the zoo cam in conjunction with the zoo visit would foster the most impact and further the skills I've asked the students to model. In future iterations, I plan to assign both so that students can further reflect on how experiential opportunities and technological advances perform different modes of reading that invariably inform one another.

Conclusion: Zoo Discomfort

This chapter has explored how the zoo, primate lab, and zoo cams can be integrated into an environmental humanities curriculum in ways that enables both public-facing and experiential humanities opportunities. I close here with a larger reflection on my own complicity of teaching with the zoo and calls for zoo abolition that have long led social justice protests for animal liberation movements. I am perpetually reminded of a satirical cutaway from *The Simpsons* in which the subtitle of the Springfield Zoo is "Come See Your Favorite Animals in Jail."

If I am honest, zoos make me uncomfortable. I am made uncomfortable by zoos because they hold immense affective resonance for me. Excitement is woven into the very fabric of the atmosphere that surrounds the zoo. I force myself to be metacognitively aware of this excitement because as a scholar of animal studies I fear the romanticization of animal innocence and cuteness, which enacts a form of social violence that reinforces and justifies human supremacy. I also fear the ephemerality of cuteness that, within dog cultures, too often exploits youthfulness, inbreeding, and physical and physiological anomalies only later to dispossess those individuals who are no longer puppies. I fear these things because they are alluring; they are hard to shake.

But this is precisely what I ask my students to do. They are, of course, excited about visits to the zoo that tear down the traditional walls of the classroom and allow us to participate in "field trips," the likes of which are limited in higher education. While I do not see it as my job to bankrupt their excitement, I do feel called upon to ask them to rigorously re-examine, as I must do every time I go, the emotional, affective, and romanticized ties we hold when visiting the zoo. I do not seek to make them uncomfortable with their feelings. I do not ask them to imagine themselves as animals—a false attempt at empathy that only effaces difference rather than attending to specificity, positionality, and preserving diverse modes of interaction, or what Gruen (2014) calls "entangled empathy." I do tell them that our zoo tourism has a purpose; it is deliberate.

Students sometimes struggle with this discussion because it forcibly pulls the veil of naïveté from over their eyes. *What do you mean all animals in the zoo are on birth control? But doesn't an elephant need more space than that? Is pacing back and forth truly a sign of stress? But aren't we extending the typical lifespan of this species by keeping it in captivity?* These are questions that invariably surface and emerge from a place of childhood fascination in which the zoo is a utopian place of wonder and excitement. My job isn't to disabuse my students of these presuppositions. My job, rather, is to invite them to think about how we can approach other questions about the zoo that truly ask what's at stake when we say "sustainability," "conservation," "quality of life," and "animal welfare." The zoo in this way is invaluable to accomplishing these larger cultural studies conversations. By shifting our discussions, students come to understand the zoo not just as a fringe entertainment anomaly but rather as a crucial site of inquiry that frames and reframes our notions of human and nonhuman animal relations.

Discomfort provides a necessary springboard for reassessing our commitments to animal captivity, all the while acknowledging that our complicity with the institution is not passive or unbeknownst to us. In venturing to the zoo, we should be made uncomfortable, and in that discomfort rises pedagogical innovation and philosophical reawakenings.

Acknowledgment

Bucknell University currently occupies the ancestral and unceded territories of the Susquehannock peoples, who remain stewards of the land, air, and water of what is now central Pennsylvania.

Notes

1 A study by Marcellini and Jenssen (1988) conducted at the reptile house US National Zoo in Washington DC found that, on average, audiences spent less than 15 minutes observing the reptile house's 74 individual exhibits, which amounts to roughly 12 seconds per exhibit. In the Lincoln Park Zoo's ape house, Ross and Lukas (2005) concluded that zoo goers spent just under 8 minutes observing six great ape exhibits—species that are housed in groups.
2 This assignment sheet can be found on the companion website.

References

ASLE, 2021. Teaching resources. Retrieved from www.asle.org/teach/teaching-resources-database (accessed October 25, 2021).

Gruen, L., 2014. *Entangled empathy: An alternative ethic to our relationships with animals*. New York: Lantern.

Marcellini, D., and Jenssen, T., 1988. Visitor behavior in the National Zoo's reptile house. *Zoo Biology*, 7, 329–338.

Naik, B., and Gumuchian, M-L., 2014. Danish zoo kills health giraffe, feeds body to lions. Retrieved from www.cnn.com/2014/02/09/world/europe/denmark-zoo-giraffe/index.html (accessed October 25, 2021).

Neimanis, A., Åsberg, C., and Hedrén, J., 2015. Four problems, four directions for environmental humanities: Toward critical posthumanities for the Anthropocene. *Ethics and Environment*, 20 (1), 67–97.

O'Gorman, E., and Gaynor, A., 2020. More-than-human histories. *Environmental History*, 25 (4), 711–735.

Peele, A., 2020. The gray cockatoo is lonely without you. *The Washington Post*. Retrieved from www.washingtonpost.com/magazine/2020/06/24/some-zoo-animals-are-doing-fine-without-visitors-many-are-lonely-without-us (accessed October 25, 2021).

Probyn-Rapsey, F., 2018. Anthropocentrism. In L. Gruen, ed. *Critical Terms for Animal Studies*. Chicago, IL: University of Chicago Press, 47–63.

Ross, S., and Lukas, K., 2005. Zoo visitor behavior at an African ape exhibit. *Visitor Studies Today*, 8 (1), 4–12.

16

PARADOX OF HOPE

Cultivating a Restorative Educational Ethic in a World on Fire

Levi Gardner

> The transition to sustainability will require learning how to recognize and resolve divergent problems, which is to say a higher level of spiritual awareness. By whatever name, something akin to spiritual renewal is the *sine qua non* of the transition to sustainability.
>
> —*Dr. David Orr (2002), "Four Challenges of Sustainability"*

My educational philosophy evolved over a decade of teaching Environmental Studies courses and countless public workshops at both university and nonprofit community farms. What began with employing predominantly depressing facts as the primary mechanism grew to responding to Orr's (2002) call for "spiritual renewal" and embracing the paradox of ecopedagogy: education in the Anthropocene can be fundamentally rooted in scientific inquiry, reason, and observation while simultaneously centered around ever-present questions of meaning and mystery. What makes the earth worth saving? What is the value of a tree? How do we live well amidst unparalleled social and ecological crises? The tension and deep friendship between the scientific and the sacred centers my teaching.

From my perspective via Environmental Studies, the entire field begins with the assumption that the world has worth and is positioned in a particularly value-laden vantage. It is the purview of work by scientists and poets, artists, and philosophers alike. Being open to such unanswerable and interdisciplinary questions meant that my teaching couldn't rely simply on data points but also had to expand outside those facts alone, ultimately prompting the question, "what is education for?"

In this essay, I explore two problems I have observed in popular ecopedagogy. These are not necessarily all-encompassing, but they do represent what I believe are fundamental shortcomings in the dominant models. The first is that while we do not lack for subject matter—from social and economic inequality to ecological collapse and systemic racism—we have few frameworks in *how* to approach these

DOI: 10.4324/9781003221807-16

issues. There is arguably so much content to explore that educators have mistakenly focused on content rather than on *how* they deliver it. Secondly, because of the increasingly personal and emotional experiences of these topics, teaching them is not simply an intellectual exercise. To do so falls drastically short. In the twenty-first century, discussing the state of the world is fundamentally as existential as it is observational. This means that our jobs are not just to inform but to guide our students—an endeavor as emotional as it is intellectual, as spiritual as it is scientific.

To that end, I attempt to distill everything I've learned about teaching into four observations that I will introduce here but expound on throughout the essay. Drawn from experience, readings, and praxis in every sense, they represent the best of what I believe is a way forward amidst these overwhelming challenges. First, the medium, which includes the instructor and the setting, arguably shapes far more about what students experience in a course than any of the content itself. What students will recall more than content is how they feel and what the setting suggested about pedagogy itself. Second, regardless of intention, outlines, syllabi, or course description, class culture always wins. No matter what sort of theorizing occurs, the proof of a praxis-oriented pedagogy is only in the shared experience of the course. Third, although students will forget most, if not all, content, their embodied experiences and memories will persist. Long after the course notes and tests have faded, students will remember how they *felt*. Our responsibility is on those experiences as much as, if not more than, the content we explore. Finally, in such an impossible and toxic socio-political climate, perhaps chief amongst all of these is the task of simply being human with and alongside our students. We must demonstrate through example the very difficult and complex task of being a human capable of complex and even paradoxical experiences. This doesn't mean the elimination of all boundaries or professionalism, but perhaps no time has ever called for us to shed any hubris that we may hold to engage the very real challenge of simply being alive in the twenty-first century.

Background

As a student turned professional turned academic turned novice artisan, I have chronicled more than a decade of teaching in various forums. That tenure witnessed multiple personal crises—familial, mental, financial, and otherwise—culminating in a PhD program that lasted one semester. Furthermore, writing this chapter coincided with the dissolution of the nonprofit farm I founded, and with it at least a suspension of my professional or academic career. Beyond that still, my age puts me squarely into the category of late millennial, so my experience with the topic of ecopedagogy is as germane to my own development as it is my teaching. I can only write from the perspective of both teacher and student, observer and participant.

I have taught thousands of students over the past decades. Pre-K through senior citizen, classes inside and outside. They have been conducted in formal college courses and informal community workshops, in hermetically sealed windowless rooms and every version of forest, field, and farm. I have built PowerPoints, administered exams, read evaluations, hosted gatherings, told recycled jokes, planned excursions, and changed direction on the fly.

What I have learned about myself, my students, and the practice of teaching could likely fill an entire volume, but I will try to articulate it here in succinct and digestible ways. Though I draw on all my diverse experiences, these observations are centered around a dozen or so semesters of teaching introductory sustainability studies courses at a Midwestern regional institution and about a decade of informal environmental education situated from educational organic farms.

How to Face the Mess We're In

During my early tenure learning about the "state of the world," nearly all books, articles, and media I consumed began with what felt like something of an assault. When introduced to the field, I inhaled books by Bill McKibben, David Orr, Bill McDonough, Paul Hawken, and a host of other predominantly white, liberal, middle-aged men. They spoke about genocide and global imperialism, the military industrial complex, and the never-ending residual of colonialism. I learned about aquifer depletion, loss of species, climate change, and eutrophication. Subconsciously my interest in the subject matter was likely inspired by my own sense of self-flagellation—something I inherited from a lifetime of shame via fundamentalist Christianity. In fact, of all the media I consumed, little aimed to encourage or instill hope. It intimidated, discouraged, and overwhelmed, but rarely was the message much beyond collapse and inescapable doom. Again, I was heavily predisposed to the narrative that "everything is going to hell in a handbasket," but I know I'm not alone in that experience.

While striding through a bookstore one day looking for yet another book on how bad things are, I discovered Joanna Macy and Chris Johnstone's *Active Hope: How to Face the Mess We're in Without Going Crazy* (Macy and Johnstone 2012). I picked it up with a bit of a chuckle; the title was so on the nose that I almost assumed it satire. This book shattered the mold I had come to know. It makes no pretense about the situation we find ourselves in but implores readers to understand and explore the connection between inner life and the world around us. The book doesn't center on the external world of action, voting, policies, or sustainability checklists, but it is an invitation to hold the impossible tension between the state of things and the gifts of joy, gratitude, and even humor. It honors the paradoxical space of both/and rather than either/or between education and activism, justice and redemption, healing both inner and outer landscapes. This book was transformational, and it, alongside a deconstruction of my fundamentalist upbringing and tiptoeing into systems theory, nonviolence, and eco-Buddhism, offered a lens to process the firehose of depressing data. It was the veritable soft shoulder rub I needed after a barrage of hellfire and brimstone.

In reflecting on what I've learned, I identify two problems that I witnessed in dominant ecopedagogy approaches. First, in most works of sustainability and social justice issues is a focus on "the mess we're in." There is a large impulse to highlight what is wrong and how bad it is, but little emphasis on how to avoid mental collapse in the face of such realities. There is even less information about how to thrive while simultaneously deconstructing systemic problems. This means that perhaps the first problem is that we (universities, scholars, and educators) have determined the "what" to teach (climate change, eutrophication, deforestation, soil erosion, colonialism, oppression and erasure of marginalized communities, eurocentrism, etc.), without determining with such ferocity the "how."

In an era replete with the dystopian realities of the COVID pandemic, George Floyd's murder, the capital insurrection, wildfires, and climate collapse as running headlines, we have few tools even remotely capable of managing this overwhelming mental burden. Arguably, if our chief aim in liberal education is to overwhelm, paralyze, and discourage the typical white, Midwestern, 19-year-old student about how they are an active participant in a white supremacist-imperialist-capitalist-patriarchy (to quote the late beloved bell hooks) and a beneficiary of centuries of exploitation and oppression, we will likely succeed. If our aim is to discourage and depress a population wherein 35 percent of undergraduate students said they "felt so depressed it was difficult to function" (Johnson and Lester 2021), we should stay the course. And if our hope is that students' desire for a better future is predominately rooted in an almost obsessive and zealous pursuit that frequently leads to burn-out, alcoholism, and depression, then this approach has in many ways succeeded. As the saying goes, the road to hell is paved with good intentions, and the intention of making students "global citizens" runs the risk of a significantly more deleterious impact.

The second, and perhaps more pressing problem, is that regardless of how it is introduced, the language of sustainability is deeper than the conversations about trees, rivers, or even the other for whom we advocate. Rather, the questions embedded in it are inherently about existence itself. We know almost unequivocally that if all humanity fades, ecosystems will shift, pioneer species will repopulate disturbed biomes, and living communities will find new equilibriums while replacing humans as the keystone predator. This is to say that when we talk about saving or protecting the world—we are speaking both of the physical world *and* of our own humanity. To explore the topics means we are explicitly discussing issues of ethics, justice, care, and love. As such, it is not only complex subject matter, but it also necessitates holding the very real spaces of anger, fear, doubt, and a host of other human experiences.

Our collective kneejerk reactions of activism in the face of global disasters often means we launch immediately into engaging solutions before simply sitting with the human experience. I recall teaching during one week of COVID via Zoom where I invited students for a time just to share their collective experiences of being students during a pandemic. Their responses existed in the emotive, not the intellectual. They expressed sadness over not sitting in a class, not moving from place to place, not getting to meet students outside of class. They talked about fear for families, uncertainty about the future, and general feelings of helplessness. The tone was a heavy one, and yet somehow (as is often the case), the grief was lessened when shared. Much later in the course, many students expressed gratitude for that time, saying that the intense politicization of the topic sometimes meant there was little space for just acknowledging complex and nonlinear experiences.

Sometimes deemed too emotive or soft, or perhaps sequestered to philosophers and ethicists, the questions faced in wrestling with the state of the world transcend the intellectual. Perhaps higher education once dwelled in a time where clear answers existed. If our postmodern era of collapse and deconstruction is revealing one thing (and it is revealing many!), the truth is that we no longer have the luxury of clear and easy solutions. As an institution, science can answer questions about the biosphere, social norms, carbon sequestration, hydrological cycles, demographic transitions, and gender stereotypes, but it does little to answer the questions of what it means to be alive—and, more specifically, what it means to live well with the full knowledge of the chaos that exists in the human condition.

These certainly aren't the only issues in environmental studies educational communities, but they are certainly present. They sit as a low vibrational bass note underneath all other discussions, and to teach as a human first and a subject-matter expert second, they fundamentally shaped my efforts. From my syllabi to course readings, class experiences to field trips, I was always aware that my goal was not subject proficiency, ecological competency, or even a primer on systems thinking. My goal was—within the best of my abilities—to facilitate the growth, encouragement, and resilience of humans who are coming of age in a remarkably complex and chaotic time.

Four Lessons

My courses covered a wide variety of issues exploring topics ranging from soil erosion to environmental justice, hydrology to urban planning. Students were generally midwestern and predominately white from a mix of conservative and liberal backgrounds. Readings included case studies and essays heavily influenced by David Orr, Macy, and Wendell Berry. After my first few semesters, I was often intentionally able to teach once weekly for three hours. The three-hour window offered me and the class a significantly improved opportunity to shape culture, experiences, and class identity than was offered in courses that met multiple times weekly.

My class itself employed various mediums and experiences: indoor and outdoor class time, imbuing class with play and art, simulated games, films, and guest speakers. As a musician and artist, I read poetry, frequently shared comedy, and invited students into brief grounding meditations before class when desired. We did stretches, played physical games, and attempted to develop a class culture that reflected the values we co-created on the first day of class. I rarely taught more than one class per semester, and all mistakes notwithstanding, I can say there are few things of which I'm prouder than those classes. In reflection, I offer four things I learned.

The Medium is the Message

Parker Palmer's book on teaching with humility, *The Courage to Teach* (Palmer 2001), should be required reading for educators of all stripes. Of his many theses, perhaps none is more pivotal than his assertion that no matter how many concepts or insights were filed away in my hippocampus and prefrontal cortex, it is ultimately the humanity and capacity for empathy of the teacher that is the true instructor. To quote the theorist Marshall McLuhan, "The medium is the message." As the medium of the content for our students, how we interact with them is the experience of the class, far more than the content around which our work orients. This lesson is immediately humbling because it necessarily means that teaching is an act of vulnerability. To encounter our shared humanity means not hiding behind intelligence, accolades, or degrees. It calls on us to explore our own humanity with our students.

Many of us can recall a particular archetypal chemistry teacher in our high school memory. He wore pleated pants and a pocket protector, told dad jokes before they were cool, and was all sorts of awkward in one-on-one conversation. But that same chemistry teacher perhaps made us excited about protons, neutrons, and atomic measurements. He was passionate about the topic, and that enthusiasm was infectious. Whether we remember the topics of the course is somewhat moot. In hindsight, we aren't necessarily able to recall any specific chemical equations, but what we absolutely recall is how we felt. We recall how he invited us to be confident, curious, or interested—skills that contribute to much deeper senses of self that far outlast our ability to balance chemical equations.

Attempting to grow as a human rather than as an educator or teacher, these words from Mr. Palmer stuck with me:

> As we learn more about who we are, we can learn techniques that reveal rather than conceal the personhood from which good teaching comes. We no longer need to use technique to mask the subjective self, as the culture of professionalism encourages us to do. Now we can use technique to manifest more fully the gift of self from which our best teaching comes.
>
> *(Palmer 2001, p. 24)*

At our core, it is our humanity and not our professional selves that dictates our students' experience. In my courses, I regularly dedicated time to the topic of pedagogy itself, and perhaps my favorite exercise was the following.

As a drummer for the better part of two decades, I would teach the course the simple pattern for eighth notes on one hand, and alternating downbeats on the other—the heartbeat of hi-hat and snare drum that pulses through nearly all western music. After a laughable class cacophony, I would invite the students to pair up and do the same. The students were allowed to pick anything they cared about, and in trading five-minute segments, each would teach the other. This often took place outside, and I saw students instruct on motions for a proper golf swing, how to cut an onion, or a simple dance move. As pedestrian as the experience was, it shattered two codified assumptions in higher education. The first is that learning is something that happens predominately when sitting down, something that I have learned to be completely untrue after years as a musician, farmer, and carpenter. The second deconstructed notion was that learning is inherently a stoic act. I saw students regularly smile and have fun, excited in the vulnerability and challenge of learning how to hold a yoga tree pose or the proper way to throw a frisbee. Perhaps it was because it was outside; perhaps it was because it was not on the test. By all accounts, I saw my students create, trust, and be vulnerable with one another—skills that again far surpass our abilities to know the right answer.

This brings me to the second part of this observation about the teacher as medium. Orr (2004) speaks of architecture as "crystallized pedagogy." I can think of few better fitting summations of the built environment, especially collegiate architecture. With narrow educational architectural debates about row seating vs. cluster seating and whiteboards vs. smart whiteboards, the assumptions underlying nearly all classrooms are indubitably the same: *Students sit, teachers stand. Learning happens indoors, mostly in the form of words. Emotion, experience, and context are subjective, whereas texts, readings, and experts hold objective knowledge. Drywall, concrete, carpet squares and other inorganic materials are preferred. Grades are the measurement of success.* I exaggerate slightly for the sake of demonstration, but in my experience, this was not an exaggeration. I taught for more than a decade in a university that boasts AASHE Silver certification, an abundant sustainability portfolio, and is regularly amongst conversations around sustainability education and advocacy. Despite all this, I regularly taught in hermetically sealed rooms with no windows, off-white walls and ceiling tiles, and neutral-blue floor tiles. If the values of the course were identified from the physical space alone, words including sterility, order, and homogeneity would have topped the list.

I won't offer an insight on the value of outdoor education; there, of course, are many works on that. I will only underscore those multiple arguments to suggest that outdoor learning is not a nice compliment, but, rather, seminal, to a humanist ecopedagogy. Limited only by daylight and temperature, I taught courses regularly outside, visiting urban rivers, gardens, rural arboretums, and city centers. Perhaps even just as important as being in the place is the spirit and culture of the class.

Culture is Co-created

In concert with understanding my role and the setting, as I learned more and more, I sought via Paulo Freire to facilitate and guide rather than to teach or educate. In its most basic form, a syllabus is a contract. It is the prescription of outlines, expectations, and the agreement between two parties. Unfortunately, in the disempowered and hierarchical structure that is the collegiate reality, many students don't know this, and many educators aren't implored to discuss it. In almost no other environment that heralds the academic values of democracy and open inquiry would one party hand a contract to the other without so much as any brief period of negotiation. In my observation, to engage students in the practice of liberatory education means to facilitate their explicit investment in shaping the process and outcomes of that education.

After going through my syllabus the first day of class, I asked what worked and what didn't. I would invite students to tell me what sounded reasonable and what didn't. What they were hoping for, why they were there, and what they hoped to gain. A relational person by nature, I always considered my first and most important contract to be with my students, my contract with the university secondary.

Perhaps never being tenured or even considered a real professor was a gift. As an adjunct, I knew full-well that my existence was a way of lowering costs for the university, so I had both little power but also few expectations. Because of that, what I learned was students deeply desired to articulate their own personal and corporate goals or even to explore whether they had them. On the first day of class I would concept map aspirational hopes the students had while keeping a running list of what was possible and what wasn't. They knew full well that we operated within limits (i.e., time, money, resources), but that we also had freedom. They asked for outside time, hands-on instruction, creativity, and guest speakers. They asked for grades to exist as guides not threats, and that study guides, cheat sheets, and notes would be liberally employed. I met these requests with my own expectations: I would give my full self; students would do the same. I would listen, challenge, and participate in the process; students would do the same. If at a point we needed to course correct, we would do so without judgement.

The famous axiom generally attributed to Peter Drucker and popular in organizational theory sometimes goes, "culture eats ideology for lunch." Regardless of theoretical frameworks, philosophies, or teaching agendas, the class culture trumps everything else. I have watched brilliant professors unable to articulate their own insecurity and fear inadvertently cultivate class cultures of fear, anxiety, and distrust.

This truth yields the largest risk and reward. It distributes power and places it into the collective, transforming class organization of teacher and student into a course facilitated by one but full of teacher-students. In this regard, the design, implementation, and co-created dynamic of the course is indubitably more important than the content itself. A course on biochemistry can leave students hopeful and excited about the world while a course on communication can take the very shape of violence.

When determining course objectives outside the codified syllabus, I tell my students I have three simple goals for them in my class: (1) learn something; (2) have fun; and (3) become healthier humans. The goals are almost pedantic in nature, yet no less valuable.

At the beginning of my last few semesters, my awareness that each semester may be my last teaching meant I was able to largely throw caution to the wind. I realize this is a particular luxury, perhaps not afforded to all educators. Regardless of job status, there is always a risk in the process of vulnerably teaching. To quote vulnerability expert, Brene Brown (2010), there is always a risk of failure when you are operating "in the arena:" the risk that you will be exposed as a fraud, that students might cheat, or that it will all go horribly wrong. Yet I believe teaching done well is more art and less science, more trial and error than formula. Brown suggests that "if we can't treat ourselves kindly, we can't practice compassion for others." What is teaching if not compassion for others?

Content Fades, Experiences are Embodied

For many years I ran a university farm. We regularly hosted classes, and when orienting students, I invited them to sit in our outdoor circle while talking about agroecology for 20 minutes before a brief tour and then a class work project. The project would vary day to day depending on the needs of the farm but was often some iteration of cultivating beds, building compost piles, sowing cover crop, or collapsing an irrigation system. For several years, these course visits were the heart and soul of the farm, synergistically pairing enthusiastic students and a physical space needing such energy.

During the last 10–15 minutes of the class, we ended with a brief activity called "Like. Learn. Notice." I truly don't recall if I made it up or read it somewhere, but it was a reflective opportunity for myself and the students. They could share something they liked *or* something they learned *or* in the spirit of nonjudgement, something they simply noticed. Without any certain program evaluation metrics, I would say approximately 5 percent of the time what the students mentioned having liked or learned happened during my introductory talk. The remaining 95 percent of the time it was what happened during their visit. "I have never actually harvested a carrot before; I liked that!" "I liked working on moving all the mulch" "I learned about how much of the soil is made up of organic matter." "I learned that hoophouses can extend the growing season." On and on it went.

Bessel Van der Kolk (2015) explains how even when we can't access particular memories, that doesn't mean our bodies can't access how we felt. In the Anthropocene and digital age, wherein we perhaps fixate on facts and data, our emotions are locked into our bodies and those experiences—both traumatic and healing—can shape our physical bodies even at a cellular level. I am no expert on these topics by any regard, but I know that when I saw students, sometimes many years later, they would light up talking about not what they *learned* but what they *felt*. As we deeply desire and long for transformational and aesthetic experiences, it is what we experience and what we feel that perhaps carries on far longer than what we learn.

Being Human Beings

Whatever our path forward, a brief glimpse of any newsreel, headline, or journal article amplifies the need for a pedagogy imbued with practices of healing, restoration, and hope, for educators to attempt to marry the sacred and the scientific, the general and the particular. In the words of the late scholar and educator E. O. Wilson (2007), "Science and religion are two of the most potent forces on Earth and they should come together to save the creation."

On my last day teaching a class on sustainable food systems, I asked my students which exam in their collegiate career had been the most meaningful to them. A consensus emerged that final examinations had primarily served as a metric of course compliance and not necessarily personal meaning. With this idea on the table, aided by my own improvisational nature, I called an audible. I offered them an option to either take a traditional content-based exam or create their own way of interacting with the class to honor their experience. I did require as a measure of safety the completion of a laborious and lengthy study guide as a price of admission. After that, all bets were off. I gave them only one parameter for their final contribution: *no judgment.* No judgment for selves, no judgment for others, no judgment for what they could or couldn't do. This included no judgment for the student that would procrastinate and offer no work. Having little idea what I was doing, I dodged questions: "So what do you expect?" "How will we be graded on this?" My aim was to invite students into owning their education rather than deferring to me—a veritable baptism into the discomfort and creativity of unknowing, one where the measurement of their education wasn't filtered through a rubric, grade, or external measuring stick. This was risky if not irresponsible, yet I was beyond confident that the students had gained a world of content in the class; I was now interested in the simple act of creation.

On the day of the exam, there was a buzz of energy in the room as none of us, myself included, knew what to expect. After deciding on an order of presentation, the first student began. An over-achiever by nature, Anna led the way. This student was one of the sharpest, most dedicated, and self-directed learners I had ever worked with. Educators can easily identify these types of students because they always ask the good questions, show up early, and turn in the highest quality work. They make teaching easy, yet often they are also the ones who struggle the most with differentiating their intrinsic value from their academic production.

She told us all how when offered the opportunity to create her own exam, her intention automatically went to "wow" us, to create something so incredible and impressive that we would all be stunned by how she captured and articulated the complexity of the class and her knowledge of it. In recognition of both the personal toll it brought on her as well as an aim to engage the process, she chose something different. Rather than produce to astonish, she opted to, during the frenetic and spastic event that is finals week, set aside time to do nothing. Instead of stressing on production, she sat quietly for 30 minutes every day during exam week. Time simply to be, not do. Her brief discussion explored how the fruit of our lives is in direct relationship to both our roots and inner life, and that capitalist production in all its forms, whether afforded by synthetic petrochemicals, dehumanizing labor practices, or self-abuse simply cannot be sustainable. In a class that had dove deep into agroecology, her words dripped of metaphor.

The next student read a letter that he wrote to the class, one he had authored to the dean of the school, and a poem by Wendell Berry. The student's bold ownership of his education, his clarity, and forthrightness of the way the class was interwoven with his Christian tradition was not only inspired, but it was entirely his. Another student who I struggled to connect with captured the spirit of the assignment effortlessly. Quite simply, he drew a picture and told a story of discovering the first urban farm he ever encountered. Still another created origami creations out of an antiquated chemistry book which she was told had no residual value.

On they went, with each student authentically sharing a part of themselves and the course content. They created and shared art, spiritual lessons, and explorations of nature and selves that simply couldn't be taught. I have had the fortunate experience to have many sacred moments in my teaching journey; that day was perhaps the apex.

Conclusion: The Gift of Paradox

Whatever the social, political, and economic institutions that inform our modern pedagogy, those same mechanisms are likely insufficient to address the complexity of teaching in the Anthropocene. Teachers (by which I mean pre-K through PhD) now more than ever occupy the space as therapists, counselors, coaches, guides, healers, and pioneers far above both their pay grades and job descriptions. Meanwhile, the existential dread, apathy, and consuming fear that has been brewing for years has only intensified.

For anyone paying attention to the world, only one choice faces us moving forward: how will we respond? How do we exist in a world on fire? Will we tuck away, numb, and succumb to the shadow? Will the doomscrolling and hopelessness seep into every corner of our existence, causing us to throw our hands up in apathy or burn with unrelenting anger? Conversely, will our privilege, titles, and professional detachment allow us to explain away the reality? Will we hide behind judgement and condescension for those who are overwhelmed because we are afforded space from it?

As is often the case, there is a third way. I believe wholeheartedly that it is the space where paradox beckons us. The space where we invite shadow and uncertainty to dine with us as a welcome guest while making room for healing, love, and curiosity at the same table. For those of us trained in western thinking, this task is not simple. It is a challenge and a gift. It is a challenge that requires us to dive into the deepest parts of ourselves—mind, body, and emotion—to entertain what appear as opposites. It's also a gift that offers us the freedom to transcend either/or realities and sets up the oddest of friends: mourning the collective loss of climate change while playing a game of duck-duck-goose. Laughing at our own privilege while honoring an overwhelming and entrenched history of systemic racism. It is where humor, gratitude, and play will be as indispensable to our path as inquiry, reason, and logic. Even as I finish these words, I sense the collective weight and uncertainty of the world hanging in the air like a mist; it is in this space our real work begins.

In the words of Wendell Berry (1989), I find almost no pronouncement as fitting for our way moving forward than his inspired language in "Manifesto: The Mad Farmer Liberation Front":

Put your faith in the two inches of humus
that will build under the trees
every thousand years.
Listen to carrion—put your ear
close, and hear the faint chattering
of the songs that are to come.

Acknowledgment

My life and work occupy the ancestral, traditional, and contemporary Lands of the Anishinabek–Three Fires Confederacy of Ojibwe, Odawa, and Potawatomi peoples. I acknowledge an entrenched historic colonialism that directly caused the forced and systematic removal of Anishinabek and other Indigenous peoples from Michigan. Furthermore, I acknowledge these transgressions and ongoing repercussions while also seeking to honor the gifts of historical and present Indigenous ways of living.

References

Berry, W., 1986. Manifesto: The Mad Farmer Liberation Front. *The Structurist*, 1989(29), 1.

Brown, B., 2010. The power of vulnerability. Retrieved from www.ted.com/talks/brene_brown_the_power_of_vulnerability?language=en (accessed January 10, 2022).

Johnson, A. P., and Lester, R. J., 2021. Mental health in academia: Hacks for cultivating and sustaining wellbeing. *American Journal of Human Biology*, e23664, 1–13. https://doi.org/10.1002/ajhb.23664.

Macy, J. R., and Johnstone, C., 2012. *Active hope: How to face the mess we're in without going crazy*. Novato, CA: New World Library.

McLuhan, M., and Gordon, W. T., 2015. *Understanding media: The extensions of man*. Cambridge, MA: MIT Press.

Orr, D. W., 2002. Four challenges of sustainability. *Conservation Biology*, 16(6), 1457–1460. https://doi.org/10.1046/j.1523-1739.2002.01668.x.

Orr, D. W., 2004. *Earth in mind: On education, environment, and the human prospect*. Washington, DC: Island Press.

Palmer, P. J., 2001. *The courage to teach: Exploring the inner landscape of a teacher's life*. San Francisco, CA: Jossey-Bass.

Van der Kolk, B., 2015. *The body keeps the score: Brain, mind and body in the healing of trauma*. New York: Penguin Books.

Wilson, E. O., 2007. *The creation: An appeal to save life on earth*. New York: Norton.

INDEX

Pages numbers in *italic* indicate illustrations, **bold** a table and n, an endnote

For Product Safety Concerns and Information please contact our EU representative GPSR@taylorandfrancis.com
Taylor & Francis Verlag GmbH, Kaufingerstraße 24, 80331 München, Germany

www.ingramcontent.com/pod-product-compliance
Ingram Content Group UK Ltd.
Pitfield, Milton Keynes, MK11 3LW, UK
UKHW020141200926
475702UK00005B/847
9781032118444